An outline of soil and rock mechanics

AN OUTLINE OF

soil and rock mechanics

PIERRE HABIB

Director, Solid Mechanics Laboratory, Ecole Polytechnique, France

Translated by Bronwen A. Rees

CAMBRIDGE UNIVERSITY PRESS

Cambridge

London New York New Rochelle

Melbourne Sydney

Published by the Press Syndicate of the University of Cambridge
The Pitt Building, Trumpington Street, Cambridge CB2 1RP
32 East 57th Street, New York, NY 10022, USA
296 Beaconsfield Parade, Middle Park, Melbourne 3206, Australia

Originally published in French as *Précis de Géotechnique*
by Dunod, Paris 1973 and

2nd French edition 1982

First published in English by Cambridge University Press
1983 as *An outline of soil and rock mechanics*

Printed in Great Britain at the Alden Press
Oxford London and Northampton

Library of Congress catalogue card number: 82-17677

British Library Cataloguing in Pulication Data

Habib, Pierre
An outline of soil and rock mechanics

1. Soil mechanics
I. Title II. Précis de géotechnique. *English*
624.1'5136 TA710

ISBN 0 521 24461 7 hard covers
ISBN 0 521 28704 9 paperback

CONTENTS

PREFACE

This book has not been written to replace the excellent treatises on soil mechanics that are already available and indeed frequently referred to throughout the text. Its aims are, paradoxically, both modest and ambitious. Modest, because it is short and the reader can, almost at one sitting gain a synthetic understanding of the basic problems created by the linear and non-linear behaviour of soil and rocks. Ambitious because in every chapter it seeks out major ideas that emerge from practical experience in an attempt to elicit a technical and scientific philosophy which I believe to be of the utmost importance, and which is without doubt far more useful than detailed description of a few isolated examples. In fact, it is clear that in confronting an unknown situation a knowledge of the various methods of approach is of greater value than is their actual resolution.

Thus, after much deliberation, I have decided to keep mathematical developments as simple as possible. Several factors led me to this decision. Long and difficult calculations of elasticity, viscoelasticity and plasticity would have rendered the general presentation tedious: the reader interested in this would do better to study the original papers, or turn to a treatise wherein the subject is comprehensively presented: a full bibliography appears at the end of the book to provide the essential references. Furthermore, mathematical rigour and the precise nature of numerical calculations create the illusion of exactitude of results while masking the imprecision of the hypothesis upon which they are founded – a notable example being the value of the mechanical coefficients of elasticity or plasticity. A great degree of uncertainty must be allowed for in determining the latter, being dependent on the very nature of things, both on the occurrence of scatter properties in natural materials as on the experimental imperfection inherent in our methods. It is not always clear what tests will provide the answers to certain questions. For example, knowing that the north face of the Grandes Jorasses is a vertical cliff of some 1200 m in height, can one determine the cohesion of a massif of granite? Or, can one find the angle of internal friction of a 10 m thick deposit of sand, under 100 m of water at the bottom of the sea? More serious, perhaps, are the questions which we believe we can resolve, but find the answers severely restricted by our means of investigation: what is the modulus of elasticity of a clay for which the cohesion is of the order of 0.2 MPa but which is fissured?

This book has been written to prepare the reader to meet just such situations. It will certainly provide satisfactory solutions to the great majority of foundation problems as it will to problems of civil engineering associated with soil or rock mechanics. But above all it is hoped that the precise description of the rheological behaviour of both soil and rocks, combined with practical examples, will ultimately lead to the right choice of hypothesis for calculating a new work, building a new construction or developing a new method of approach.

PREFACE TO THE ENGLISH EDITION

Nowadays problems relating to the stability of foundations, and to the equilibrium of large earth or rock masses are becoming ever more important. This is due to the increasing size of major engineering works such as: high rise buildings, deep excavations for underground garages, offshore oil rigs, underground openings for subway or motorway tunnels.

The diversity of techniques used in the construction of such works has created the need for a short yet comprehensive book outlining the fundamental principles used by the specialist. Such a book should enable the engineer quickly to acquire a thorough knowledge of the principles, and to apply these with sufficient precision to a given problem.

Such is the aim of this book and the warm welcome extended to it in France demonstrates the value of such an approach.

P. Habib 1982

UNITS

The units used in this book are those of the Système International and their multiples. The basic units and main magnitudes used in mechanics are:

Mass:	kilogram (kg)
Force and weight:	newton (N): 1 N ≡ 0.102 kgf
Length:	metre (m)
Stress and pressure:	pascal (Pa): 1 Pa = 1 N/m^2
Density:	kilogram per cubic metre (kg/m^3)
Unit weight:	newton per cubic metre (N/m^3)
Time:	second (s)
Permeability:	metre per second (m/s)
Viscosity:	pascal second or poiseuille (Pl)

1 Pa × s = 1 poiseuille = 10 poise = 1 Pl = 10 Po.

It is suggested that the different magnitudes be written:

- Either by using the accepted units as above preceded by power of 10. In this case it is advisable to choose this power so that the number used has only one significant figure to the left of the decimal point.
  For example:
  $k = 3.5 \times 10^{-4}$ m/s and not $k = 0.035 \times 10^{-2}$ m/s
  and not $k = 350 \times 10^{-6}$ m/s
  $\gamma = 1.92 \times 10^4$ N/m^3 and not 192×10^2 N/m^3
  $\gamma_s = 2.70 \times 10^4$ N/m^3 and not 2700×10^1 N/m^3.
- Or by using the accepted multiples: deca, hecto, kilo, mega, giga.
  For instance:
  1 kilonewton = 1 kN = 102 kgf
  1 kilonewton per cubic metre = 1 kN/m^3 ≡ 0.102 tf/m^3 ≡ 102 kgf/m^3
  1 megapascal ≡ 1 MPa = 10 bar ≡ 10.2 kgf/cm^3
  1 gigapascal = 1 GPa = 10 kbar
  η (water) = 1 millipoiseuille (1 centipoise).

1

INTRODUCTION TO SOIL MECHANICS – SOIL FOR THE ENGINEER

1.1 Introduction

While soil mechanics resembles and indeed derives many of its features from the mechanics of continua, its uniqueness lies in the granular nature of its constituents. Although at the microscopic level soil is a discontinuous medium, the number of its constituents is such that more often than not the concept of continuity can be maintained: to give an order of magnitude, a thimble full of fine sand would contain about a million grains. The porosity associated with the granular structure corresponds to the free volume between the grains. This volume is full of fluid, liquid or gas, water or air. Deformation is thus accompanied by fluid pressure resulting in hydrodynamic movements.

Contact between the grains is made either at points, as in soil, or at faces, as in rocks. As a result of external forces stresses develop between grains, the contact areas increase, and the soil undergoes non-linear deformation. When the stresses increase, slipping takes place and the initial structure of the soil is destroyed: the limit of resistance has been reached. Soil mechanics is the study of the soil's properties of resistance and deformation for application in construction work.

1.2 Investigation in soil mechanics

After morphological, geological and hydrological examination of the site (the scope of which does not enter into this book) research should be carried out to determine the mechanical characteristics of the soil before making plans for foundations. The considerable impact that the nature of the earth and the type of foundation can have on the cost of a project necessitates that the correct choice be made at this stage to avoid costly modifications during construction. Two cases can be distinguished, and are discussed in sections 1.2.1 and 1.2.2.

1.2.1 *Choosing the type of foundation for a construction*

In this case the investigation is carried out by trial borings or *in situ* tests at points on a grid covering the zone. Since the investigation is only conducted at points, one must interpolate between the points. The examination is thus, by its very nature, incomplete, but bearing in mind the high cost of trial boring it is seldom commercially viable to apply stricter standards. The decision to halt the operation should be taken as soon as it is apparent that further searches will yield little supplementary information. If, however, there is any suspicion of a danger point then the examination must be continued.

1.2.2 *Choosing earth for embankments*

First, suitable local materials should be sought to provide the most stable embankment in the most economic manner. This means that the volume of earth locally available should be evaluated as well as its quality. Also, in this case, the 'pin prick' investigation of the type described in section 1.2.1 must be accompanied by linear examination, surface trenches, pits, or adits.

Once the broader categories of soil have been recognised the laboratory examines samples either to identify them or to determine their mechanical properties.

1.3 Identification of soils

The problem of soil identification lies in characterising a material sufficiently clearly so that it can be meaningfully compared with other materials bearing similar constructions or compared with different states of the same material. Besides the immediate identification (colour, smell, texture) a series of laboratory tests exists which permits a high degree of precision.

1.3.1 *Grain size analysis*

The sieving method is used for particles larger than 80 μm. Finer grains are analysed by sedimentation in a liquid (wet analysis) making use of Stokes' law and assuming the particles to be spherical. The type of sieve adopted in the French standard is defined by mesh openings which follow Renard's series $\sqrt[10]{10}$, which makes it easy to translate from sieves (square openings) to strainers (round openings). But the grain size distributions illustrated in fig. 1.1 are only significant statistically: given that the particles have a 'normal' shape, the same weight of cohesionless product would pass through a sieve with a mesh opening width $2a$ as would pass through a strainer with a hole diameter $2a\sqrt[10]{10}$. This correspondence would obviously not hold for particles with too regular a shape, such as spheres, or too eccentric, such as needle-like grains.

Sedimentation is, like sieving, a conventional test which should be carried out in strictly standardised conditions. It must be remembered that the method of separating the particles (be it grinding by dry mortar, slaking, pneumatic or mechanical agitation of suspensions, etc.) or the maintenance of that separation

Fig. 1.1. Grain size analysis of some soils. The soil categories are shown according to their grain size. The different horizontal scales demonstrate the correspondence between the sieves of the American standard ASTM, the mesh of the sieves, the diameter of the round-hole screen and the French standard AFNOR.

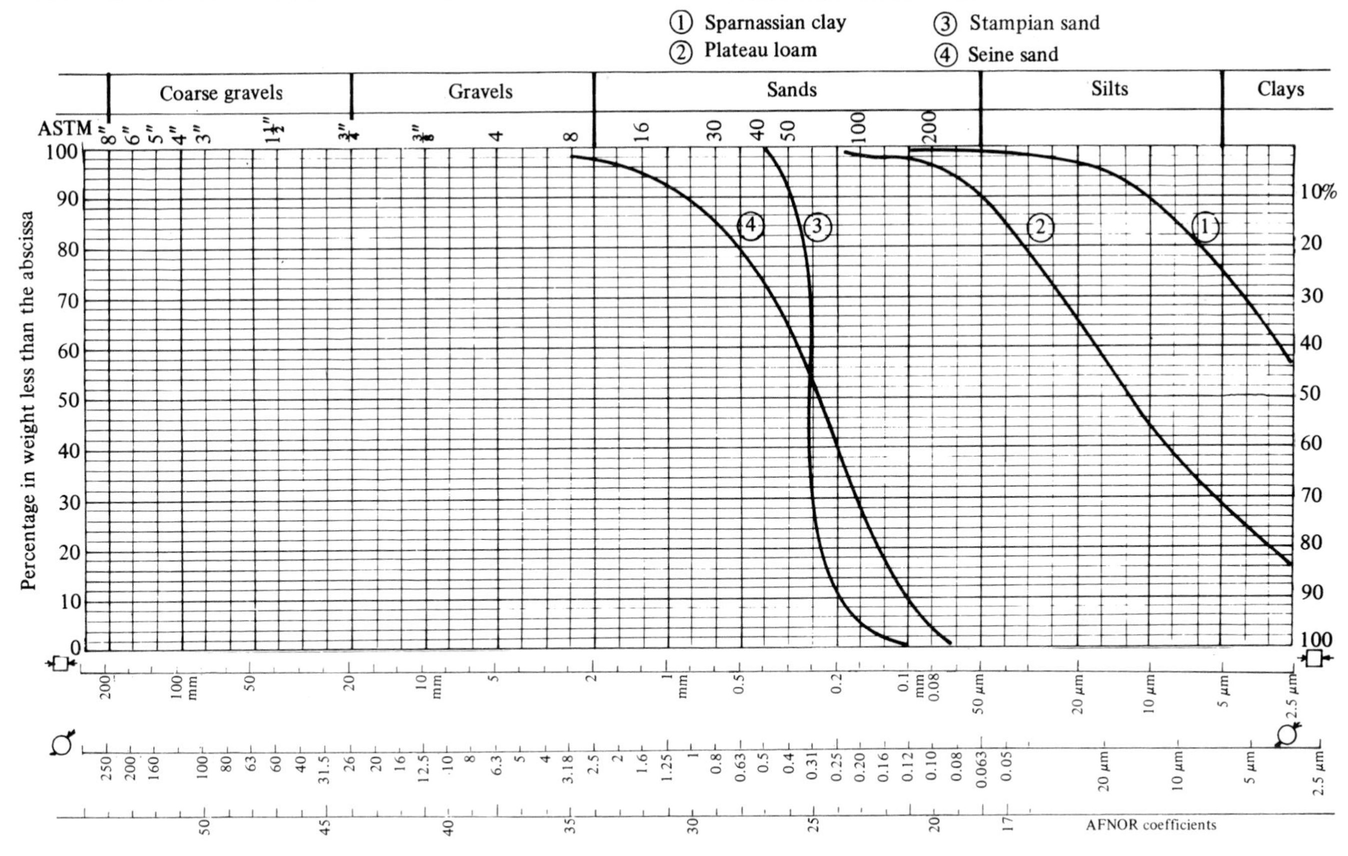

(mechanical aids in sieving, pH of suspensions, use of deflocculants, etc.) completely condition the results. Then when results from different laboratories are compared it is a necessity to be sure that the experimental conditions are identical.

1.3.2 *Natural water content – organic matter*

The natural water content $w\%$ is the ratio of the water in the soil (conventionally defined by the loss of weight after 24 hours in an oven at 106 °C) to the weight of solid matter in the soil. In certain cases the loss of weight in the oven may be due to the dehydration of certain crystals, notably gypsum, or to the destruction of organic matter. If this is likely, chemical analysis may be used to reveal the presence of unstable, hydrated crystals, which may then be independently separated from the organic matter – for example by destroying them with hydrogen peroxide and measuring the corresponding loss in weight.

The water content of soils is extremely variable. In clays, for example, the water content is typically between 30 and 60%. However, very stiff, strongly consolidated clays can have $w = 10\%$ and very soft colloidal clays can have $w = 200\%$. In sands the extreme values are much closer, typically from 15 to 35%; on the other hand they are sometimes unsaturated.

1.3.3 *The Atterberg limits*

The Atterberg limits are the conventional limits of water content which define the state of a soil. Tests to determine these limits are carried out on the fraction of earth that passes through a 0.5 mm undersize sieve. By definition the liquid limit ($w_L\%$) is the water content above which the soil behaves like a semi-liquid and flows under its own weight. The plasticity limit ($w_P\%$) is the water content under which the soil loses its plasticity and becomes friable. Conventional tests are carried out to determine these states; for instance if the soil can be manually formed into a roll 3 mm in diameter it is said to be in the plastic state. One might find the crude nature of such tests somewhat amusing, but they correspond perfectly to changes in mechanical behaviour, and, when carried out in standardised conditions, are perfectly reproducible. Comparison between the natural water content and the Atterberg limits provide an immediate idea of the actual consistency of the soil. The difference between the liquid limit and plasticity limit ($w_L - w_P$) is known as the plasticity index (PI). This will be all the higher if the soil contains a clay. A soil having PI > 10% is fairly clayey, and for PI > 30% it is very clayey.

It is sometimes found useful to define the shrinkage limit (w_R) which is the water content below which no further shrinkage takes place during desiccation: all the particles of the mineral skeleton are then in perfect contact. This is, unfortunately, less precise than the limits defined above as it depends on the initial water content of the soil when it is set out to dry, and on its state of saturation at the beginning of the test.

1.3.4 *Void ratio – porosity*

The voids in soil are not necessarily filled with water, so that the compactness of the granular structure has to be defined in terms of convenient parameters. The void ratio e is the ratio of the volume of voids to the volume of solids. Porosity n is the ratio of the volume of voids to the total volume ($e = n/(1-n)$). The degree of saturation $S\%$ of a soil is the ratio between the volume effectively taken up by the water and the total volume of the voids. The water content of a saturated soil is denoted w_s (fig. 1.2).

1.3.5 *Density – specific gravity*

International usage has brought with it a rather incorrect terminology for these expressions. Generally in Statics, one studies forces, therefore weights, and densities. In Soil Mechanics however the term density is used in the following, easily understandable terms. The symbols and meanings are indicated after the term:

bulk density, γ (weight of unit volume of moist soil); dry density, γ_d (weight of unit volume of soil, excluding water); specific gravity, γ_s (weight of unit volume of solid particle); submerged density, γ_i (weight of unit volume of soil by Archimedes principle); liquid density, γ_w (weight of unit volume of pore liquid).

Fig. 1.2. Composition of a unit volume of soil showing the relationship between solid, liquid and gas.

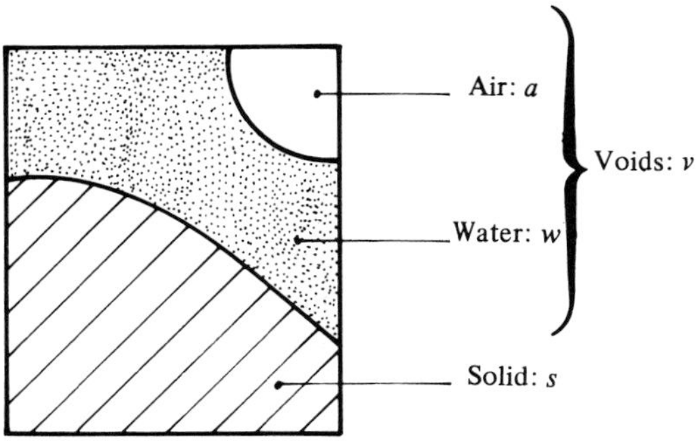

Air: a
Water: w
Voids: v
Solid: s

$$s + v = s + a + w = 1$$

$$n = \frac{v}{v+s} = v = \frac{e}{1+e}$$

$$s = 1 - n = \frac{1}{1+e}$$

$$e = \frac{v}{s} = \frac{v}{1-v} = \frac{n}{1-n}$$

$$\bar{a} = \frac{a}{a+w}$$

$$S = \frac{w}{a+w} = \frac{w}{w_s}$$

One can immediately find the following relations:

$$\gamma = \gamma_d(1+w),$$
$$e = w_s\gamma_s/\gamma_w,$$
$$w_s = \gamma_w[(1/\gamma_d)-(1/\gamma_s)],$$
$$e = (\gamma_s/\gamma_d)-1,$$
$$\gamma = (1-n)\gamma_s + n\gamma_w,$$
$$\gamma_d = (1-n)\gamma_s,$$
$$\gamma_i = (1-n)(\gamma_s-\gamma_w) = \gamma-\gamma_w.$$

Let us give some numerical values for density, for example; γ can vary from 1.7 to 2.2; γ_d is a fictitious quantity of the order of 1.4 and less in clays, and 1.6 and more in sands; γ_s is always around 2.7 in clays, calcite, quartz and feldspar, in other words in practically all inert constituents of soil; γ_i is in the neighbourhood of unity (from 0.7 to 1.2); γ_w is equal to 1 in water but for certain geotechnical problems pertaining to oil $\gamma_w = 0.9$, and $\gamma_w = 1.3$ when the interstitial fluid is a highly salinated water. Table 1.1 gives the saturated water content w_s as a function of dry density γ_d for grains whose specific gravity has the value 2.70 g/cm^3.

In the case of sands, the idea of relative density D_r is used to determine their actual compactness,

$$D_r = \frac{e_m - e}{e_m - e_M} = \frac{\gamma_{dM}}{\gamma_d}\frac{\gamma_d-\gamma_{dm}}{\gamma_{dM}-\gamma_{dm}},$$

Table 1.1. *Saturated water content as a function of dry density (specific gravity of particles $\gamma_s = 2.70$ g/cm^3).*

γ_d	w_s%	γ_d	w_s%	γ_d	w_s%	γ_d	w_s%
1.30	39.89	1.50	29.63	1.70	21.79	1.90	15.59
1.32	38.72	1.52	28.75	1.72	21.10	1.92	15.05
1.34	37.59	1.54	27.90	1.74	20.43	1.94	14.51
1.36	36.49	1.56	27.07	1.76	19.78	1.96	13.98
1.38	35.43	1.58	26.25	1.78	19.14	1.98	13.47
1.40	34.39	1.60	25.46	1.80	18.52	2.00	12.96
1.42	33.39	1.62	24.69	1.82	17.91	2.02	12.47
1.44	32.41	1.64	23.94	1.84	17.31	2.04	11.98
1.46	31.46	1.66	23.20	1.86	16.73	2.06	11.51
1.48	30.53	1.68	22.49	1.88	16.15	2.08	11.04

where the indices m and M correspond to the minimum or maximum compactness in a particular standardised system (A.S.T.M., 1946). Unfortunately the determination of γ_{dm} and γ_{dM} is very sensitive to experimental conditions and hence the value of D_r is not always very precise. However it does mean that sands can be classified into very loose, loose, medium, compact and very compact, thereby serving as a useful guide in comparing different materials.

1.4 From identification to classification

The results of the tests described in section 1.3 will furnish sufficiently precise information to enable the categories of soils and their possible areas of use to be defined. Various classification systems have been proposed, where the variations between systems are due essentially to local differences between soils (Casagrande's classification, 1947, in France, Classification L.C.P.C., etc.).

Nevertheless it is advisable to consult two diagrams, that of Taylor (1956) based on particle size analysis (fig. 1.3), and that of Casagrande, based upon the Atterberg limits (fig. 1.4). Natural clays are a mixture in various proportions of perfectly defined mineralogical varieties: kaolinite, illite, halloysite, montmorillonite, etc. The identification of soils by the geotechnical examination described in section 1.3 is certainly very important, but mineralogical classification should not be neglected.

Fig. 1.3. Taylor's diagram. The following granular classes are shown in Taylor's diagram, after elements larger than 2 mm have been eliminated: Sand, between 2 mm and 50 μm; Silt, between 50 μm and 5 μm; Clay less than 5 μm.

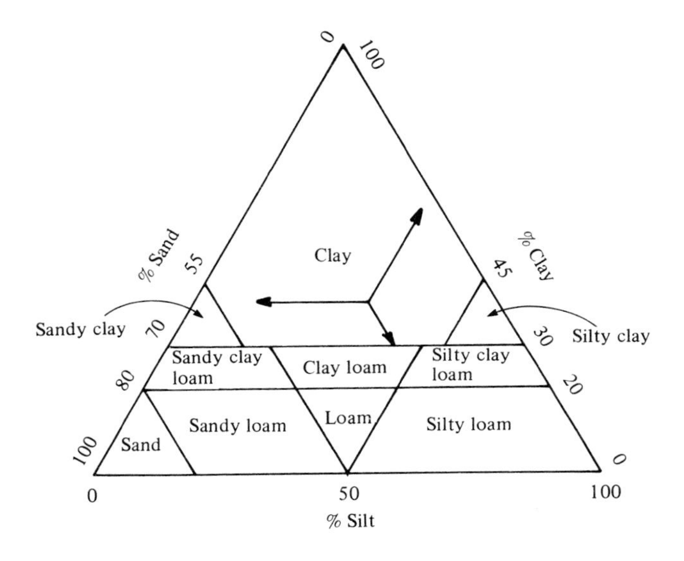

It is a well-known fact for example that as a result of its affinity for water, montmorillonite is a dangerous soil, capable of extensive swelling with a spectacular loss of resistance. However, its presence would not pass undetected: the natural water content is always very high, as are the Atterberg limits ($w_L = 400\%$ in montmorillonite, $w_L = 600\%$ in pure bentonite!). On the other hand, halloysite, which has almost as marked a tendency to lose its resistance, has no notable geotechnical properties, in particular none relating to its water content, which enable it to be identified easily. Although intrinsically less dangerous than montmorillonite, halloysite has, paradoxically, been the cause of many more accidents. It is thus highly desirable that mineralogical analysis is used to determine the nature and composition of clays.

1.5 Permeability

Imagine a cylindrical shell filled with soil with pressurised water entering at one end and emerging at the other. Such tubes, called piezometers, allow the pressure of pore water to be measured at different points (fig. 1.5). The pressure of water in piezometers decreases linearly with distance whatever the feed pressure. The head losses are proportional to the thickness of earth crossed since circulation in the soil is laminar.

Let h be the head of water expressed as the height of water and s the distance traversed. The hydraulic gradient is expressed in the relation $i = h/s$ and the rate of percolation is proportional to i. The relation between hydraulic gradient and the rate of percolation is very important in soil mechanics and is known as Darcy's law: it is usually expressed in the form

$$V = ki,$$

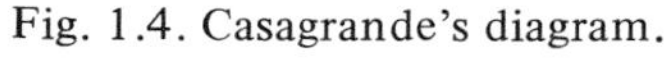

Fig. 1.4. Casagrande's diagram.

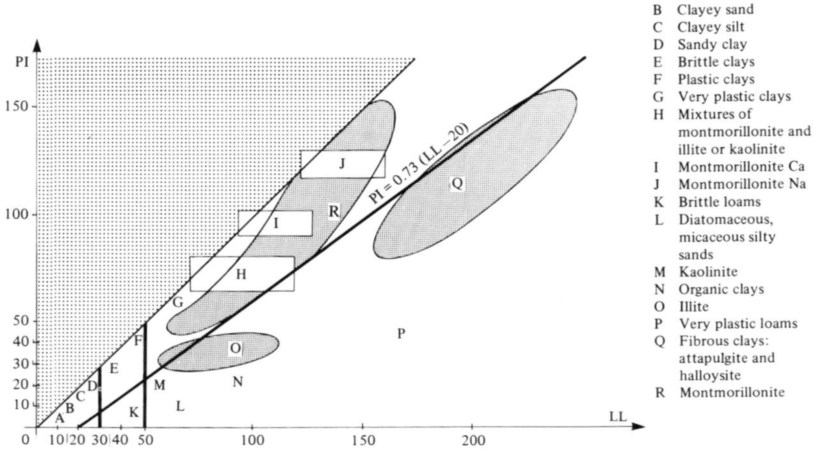

where k is a constant determined by the nature of the soil called the coefficient of permeability, and V is the velocity of water exterior to the soil sample.

The rate of discharge Q in a time Δt in a tube of cross-section S can thus be expressed

$$Q = kiS\Delta t.$$

In soil the average velocity is V/n since water only circulates through the pores. The real velocity of water is somewhat greater, and some dispersion takes place, for the capillary pathways are neither rectilinear nor uniform.

If, in Darcy's law, h is expressed in metres of water and s in metres, i is dimensionless and k has the dimensions of a velocity. This dimensional formulation is not very satisfactory for the physicist, particularly since the viscosity of water is implicit in the constant k and this in turn depends on the temperature. However it is extremely convenient for engineers.

Soil permeability is closely linked with the dimensions of the pores, in other words with the grain size of the particles and their state of compactness. In coarse soils such as sands a rapid approximation can be reached by using the (non-homogeneous) Hazen's formula

$$k\ (\text{cm/s}) = 100D_{10}^2\ (\text{cm}),$$

where D_{10} represents the diameter of grain such that 10% by weight of the soil is finer. The coefficient k varies enormously from one soil to another. For example k can lie between 10^{-2} and 10^{-6} m/s in sands, between 10^{-7} and 10^{-9} m/s in loams, between 10^{-9} and 10^{-12} m/s in clays and between 10^{-9} and 10^{-16} m/s in rocks. In the laboratory the coefficient k is measured by a prolonged percolation test. Given the range of values indicated above, it is necessary to have several types of permeameter available. The smaller k, the smaller the measured flow, the larger the hydraulic gradient and the longer the duration of the test.

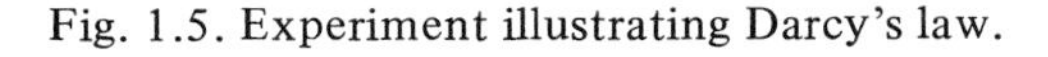
Fig. 1.5. Experiment illustrating Darcy's law.

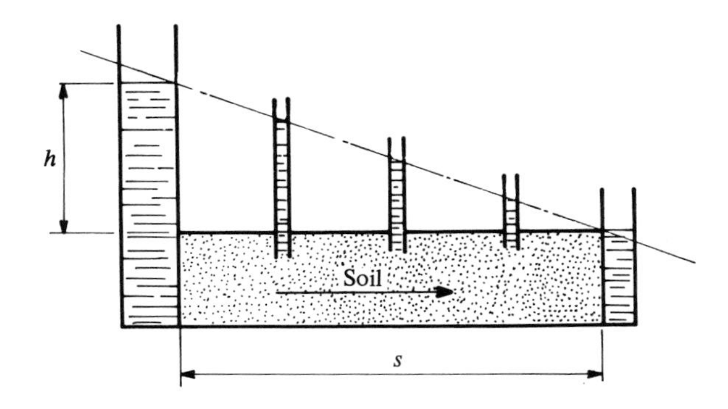

Finally, it must be remembered that these are only 'small-scale' tests, carried out on samples. In nature, other flow paths can occur (stratification joints, fissures, conduits, etc.) which must be studied by *in situ* tests. If there is any possibility of problems with seepage, then 'large-scale' tests of permeability *in situ*, such as pumping or injection into the water table, are essential.

1.6 Structural analysis of soils

That solid is not the homogeneous and isotropic body encountered in rational mechanics is well illustrated by the differences found in permeability on a 'large' and 'small' scale. The majority of mechanical properties are determined by the texture of soils (Rowe, 1972), that is by the shape, size and arrangement of the solid particles, organic inclusions, and the pores. One sometimes encounters in nature certain structures which disrupt the textural arrangement of a soil. This, for example, is what happens when silt beds are found in the interior of inter-stratified clays (varved clays): in this case numerous deposits, whether quaternary, river or glacial are formed following the annual cycles. Sedimentation of clay and fines takes place during one season, and sand and silts the next. The permeability can vary from 1 to 100 or from 1 to 1000 depending on the direction of percolation. Obviously a homogeneous mixture of all the particles would give an entirely different product, particularly from the point of view of permeability. The anisotropy due to stratification is the clearest example of the importance of grain architecture.

The identification tests described in section 1.3 do not take into account the position of the particles relative to one another. They are deemed to be adequate when carried out on complete soil samples – that is samples where no granular class has been neglected during the taking of the sample (for example by being washed out). The structure of the soil can only be found by carrying out tests on undisturbed, or representative samples – that is samples which have conserved intact the architecture of the soil. This type of investigation is similar to those carried out by petrologists or mineralogists (observation, photographing and enlarging cuts and fractures, granular analysis, analysis of the interstitial products, describing the spatial lay-out). Such research is aimed at defining two heterogeneous types: first, the stratification anisotropy whose role in permeability has already been outlined, but which also plays a part in the mechanical properties of deformation and resistance; and second, natural fissuration detected for example in clays by the slipping marks (slickenside) that manifest themselves in the fracture of samples during desiccation; the permeability of these fissures (that is the granular structure of the filling material, if there is any) provides important data for the structural analysis. A knowledge of the range of structural heterogeneities allows one to work out the minimum dimensions of the test samples necessary to measure the mechanical properties studied in chapter 2.

2

RHEOLOGICAL BEHAVIOUR OF SOILS

2.1 Shearing resistance

2.1.1 *Coulomb's law*

Under excessive external forces, soil undergoes plastic deformation and slipping occurs. The shear strength of a cohesionless material is determined by the contact forces and friction between the grains; experiments show that slipping occurs on a surface when the stress acting on it has reached an angle φ to the normal at the surface. Thus at the moment when slipping begins

$$\sigma_t = \sigma_n \tan \varphi,$$

the form of which is identical to the friction law;[1] σ_t is the shear stress required for failure to occur, known as the shearing resistance, σ_n is the stress normal to the slipping plane and φ is called the angle of internal friction. In sand the angle φ depends essentially on the compactness of the medium. In granular materials experiments give a relation of the form

$$e \tan \varphi = C,$$

where e is the void ratio and C is a constant of the order 0.45 to 0.55 depending on the nature of the particles, their shape, surface state and granulometric distribution. However, in cohesive materials, such as clays and loams, physical bonds exist between the particles in addition to the contact forces. The equilibrium conditions at slipping become:

$$\sigma_t = c + \sigma_n \tan \varphi,$$

where c is the cohesion. This is Coulomb's law.

It is not necessary to attach too precise a physical significance to c and φ since this would not stand up to detailed analysis, they should be treated instead as two coefficients of a linear approximation; adjustment of the coefficients for closer agreement with Coulomb's law is best done using experimentally derived data.

In the criteria for failure outlined above, it is assumed that the stresses are

1 In soil and rock mechanics, and civil engineering in general it is usual to adopt the sign convention that compressions are positive.

acting on the particles. In reality, as there is water in clay, the application of σ_n results in the appearance of a pore pressure u. As the shearing resistance of water is zero, Coulomb's law should be written:

$$\sigma_t = c + (\sigma_n - u) \tan \varphi$$

This is known as Hvorslev's law.

2.1.2 *Measuring shearing resistance*

Two types of test exist to measure shearing resistance – the direct shear test and the triaxial test.

Of the different direct shear test apparatus available, the best known is that of Casagrande. The soil sample is enclosed in a split-box, the top half of which can slide over the bottom half (fig. 2.1*a*). A load is applied by a piston normal to the slip plane and the shearing stress required to produce failure is determined. σ_n and σ_t values at failure are plotted on a graph as shown in fig. 2.1*b*. The pair of values (σ_n, σ_t) define a point on a Mohr diagram. The intrinsic curve, which according to Coulomb's law is a straight line, shows the amount of stress the soil can bear without failure.

In the triaxial test a cylindrical sample contained in a thin rubber membrane is laterally submitted to fluid pressure ($\sigma_2 = \sigma_3$) and axially to pressure from a piston, (σ_1), (fig. 2.2*a*). Failure graph (fig. 2.2*b*). Failure manifests itself either by excessive deformation or by a slip plane. Mohr's circle can then be drawn in ultimate equilibrium. The envelope of Mohr's circles is the intrinsic curve, which, in soils, is Coulomb's line.

For cohesionless materials the intrinsic curve is in the main rectilinear. However, under large stresses (of the order of MPa) in a river sand such as that of the Seine, attrition (granular destruction) is likely to occur. The slope of the intrinsic curve simultaneously diminishes and begins to take on a parabolic form. In a homogeneous material, the smaller the particles the greater the stress causing

Fig. 2.1. (*a*) Schematic diagram of apparatus used for the direct shear test. (*b*) The intrinsic curve.

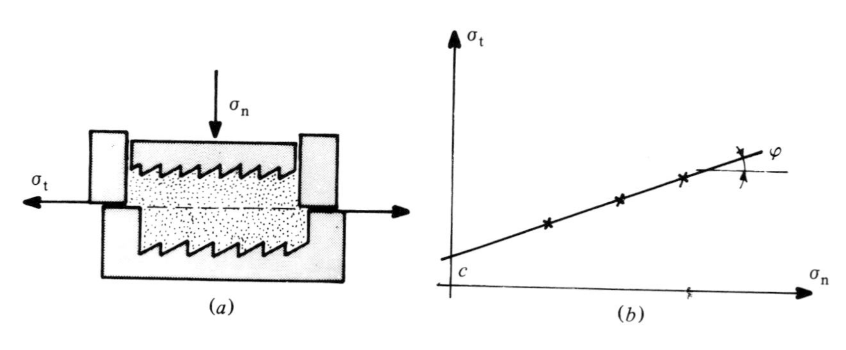

the attrition. Only at very high pressure of some 20 MPa would an analogous intrinsic curve be observed for clays but the particles would probably not be destroyed.

2.1.3 *Volume variation during shearing*

Sands

Grains of sand separate when slipping resulting in variations in volume which are determined by the initial density: dense sands dilate; loose sands become compressed. Critical density γ_c is the point at which no further volume change takes place during shear. It is a function of σ_n. After significant slipping, volume variation ceases and the corresponding shearing resistance is called critical resistance (critical angle of friction: φ_c). In sands whose initial density is greater than the critical density, there is a peak in the stress/strain curve (fig. 2.3). In the reverse case, the curve is flatter than that for critical shearing.

Fig. 2.2. (*a*) Principle used for the triaxial test. (*b*) The intrinsic curve is the envelope of the Mohr's circles determined by the test.

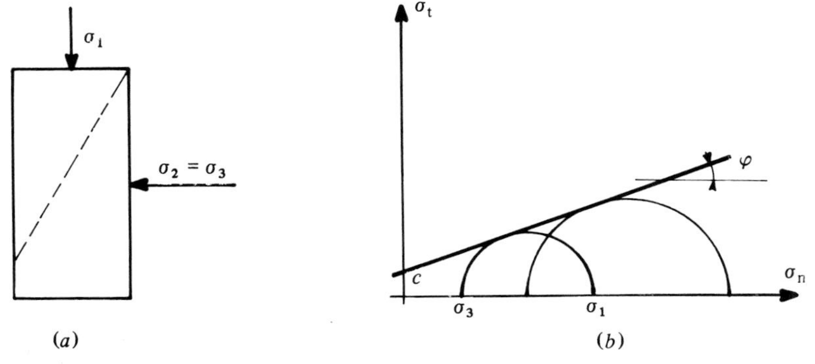

Fig. 2.3. Shearing resistance σ_t as a function of the displacement during shear tests on sands with different densities.

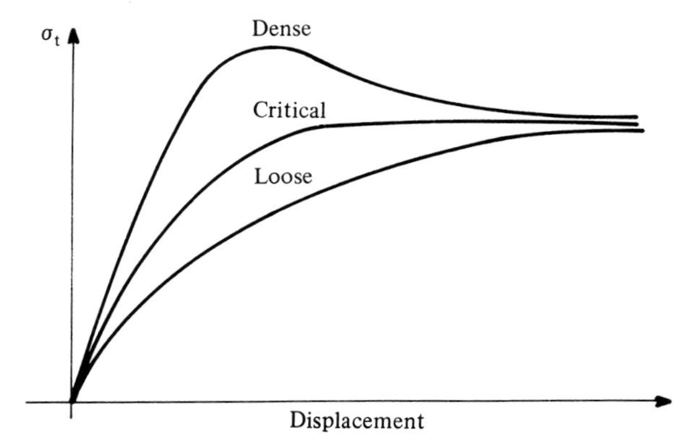

Such diagrams are very characteristic; they illustrate the relation between φ and the compactness discussed earlier.

The angle of internal friction at or around critical density is often of the order of 32°.

Clays

The granular skeleton of clays often has a density less than the critical density. However, the presence of water prevents volume reduction and instead, during the test, pore pressure appears, denoted u. This can be measured as it develops and the test can be intepreted in terms of effective stresses σ_t and $(\sigma_n - u)$ or $(\sigma_3 - u)$ and $(\sigma_1 - u)$, but it is important to differentiate this pore pressure from that mentioned earlier.

In fact, experimental conditions must be considerably modified to allow for different drainage conditions during testing, and the following tests should be distinguished:

consolidated, slow (dissipation of pore pressure) (φ', c');

consolidated, quick (no dissipation of pore pressure that arises during shearing, but previously drained of pore pressure that arises when subjected to normal stress) (φ_{cu}, c_{cu});

unconsolidated, quick, or undrained test (no dissipation of pore pressure) (φ_u, c_u).

The three tests give different Coulomb lines. In particular the unconsolidated quick test on saturated clays of low permeability gives an apparent angle of internal friction of zero. In the above definitions the work 'quick' should be taken in a sense relative to the permeability of the soil, size of the sample and to the method of testing.

Even though the 'quick' shear tests are usually carried out on clays at constant volume, the stress/strain curves have an analogous shape to those in fig. 2.3, and are functions of the initial density of the soil. The appearance of a peak in certain cases is understood to be the destruction of a structure in the slip plane.

2.2 Compressibility

2.2.1 *Consolidation of soils*

When soil is subjected to a uniform pressure the granular skeleton undergoes non-linear deformation (Hertz contact).

If the soil is initially saturated, the uniform pressure at the moment of deformation is almost completely transferred to the pore water. Since the compressibility of the water (and of the grains) is incomparably less than that of the solid skeleton, water is progressively expulsed, the volume decreases and the stresses act once more upon the grains: this phenomenon is known as settlement. The more impervious the soil, the slower equilibrium is reached. When the stresses act only on the solid skeleton ($u = 0$) the soil is said to be consolidated.

2.2.2 *Measuring compressibility*

The chief components of the apparatus used to measure compressibility, the oedometer, are a piston and a cylinder (fig. 2.4*a*). The soil sample is compressed between two porous stones through which water can escape. Measurements are taken of the settlement under load and the swelling at the moment the load is removed, and results are plotted on a void ratio versus log pressure graph which gives a convenient representation (fig. 2.4*b*).

Natural soil is most often in equilibrium under a weight of earth γh, where γ is the bulk density and h is the height of earth. If a natural sample is taken without destroying the structure of the soil, a diagram can be drawn to find the consolidation pressure p_c of the soil, or, in other words, the load to which the mineral skeleton of the soil was subjected. If $p_c = \gamma h$ then the soil is said to be normally consolidated; if $p_c > \gamma h$ the soil is over-consolidated and finally if $p_c < \gamma h$ the soil is under-consolidated.

The oedometer can in principle also be used to measure permeability. This should be done after the soil has been placed in hydrodynamic equilibrium under a load p, and the permeability variations measured as a function of soil compaction.

If the soil is at all irregular, stratified or fissured, then such a method of measuring permeability may yield different results for small samples than those measured in larger volumes, as for example in tests that may take place *in situ*.

2.2.3 *Secondary settlement*

In very soft soils hydrodynamic consolidation (which tends asymptotically towards finite deformation) is accompanied by a delayed deformation of a different nature that follows a linear law in log time. This complementary settlement is negligeable in clays and old loams, but it can be

Fig. 2.4. (*a*) Principle of the oedometer test. (*b*) The compressibility diagram.

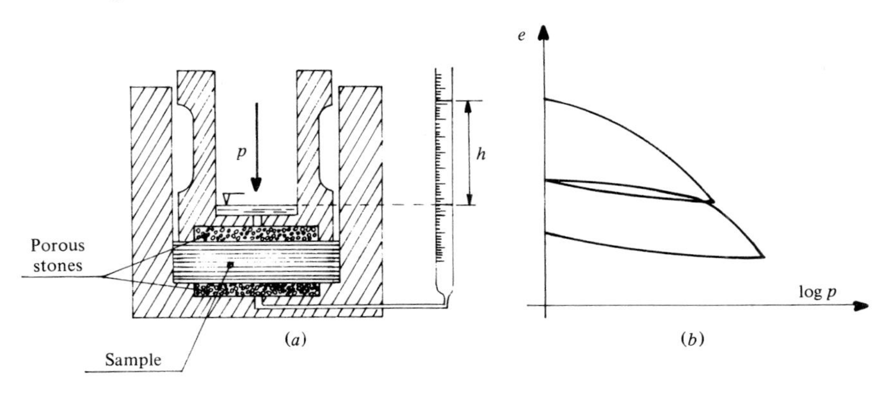

significant in recently formed soils such as muds and peats. It is called secondary settlement and only becomes apparent after hydrodynamic consolidation. Since the expulsion of water is very slow and the pore pressure very low the rate of settlement is just about independent of the sample size.

2.3 Relation between shearing strength and compressibility

Close examination of the results of a shear test on a naturally consolidated sample will reveal that the intrinsic curve is not rectilinear but, on one side of the consolidation pressure is a straight line which would pass through the origin, and on the other side of the consolidation pressure is a curve lying above this straight line (fig. 2.5).

This result can immediately be linked to the hysteresis of the consolidation diagrams.

Coulomb's law thus appears to be an approximation and is valid only within a well-defined region.

2.4 Sample taking for mechanical tests

The sampling and transport of soils are delicate operations which require experience and care. When the samples reach the laboratory they should not only be complete and representative but they should still retain their natural density and structure.

The quality of the measurements of the mechanical characteristics depends on the quality of the samples tested, and at a certain stage all soil mechanics use numerical values for resistance (φ and c) or compressibility. Now, except in exceptional cases (visibly deformed stratifications, irregular inclusions), it is practically impossible to recognise in the laboratory whether a sample is complete (e.g. without loss of water), representative (e.g. a structure may have been damaged by frost whilst being transported by plane), or intact (e.g. a

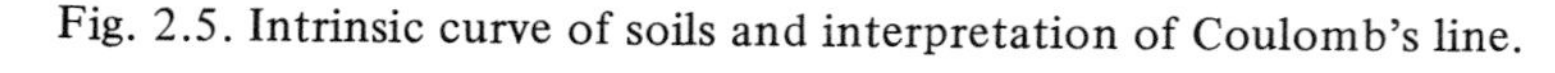

Fig. 2.5. Intrinsic curve of soils and interpretation of Coulomb's line.

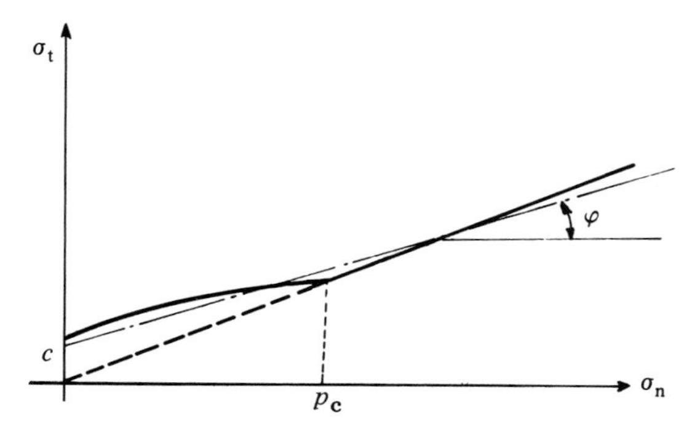

cohesionless sample may have been compacted during sampling). Samples should thus only be taken under the closest supervision. It should be particularly noted that, in arranging a working season of trial borings, it is dangerous to spend too long looking for the most competitive tender, since the quality of samples, essential for a precise study of the soil, may suffer.

Intact samples can either be taken by hand from a shaft or excavated from a trial borehole. In the latter instance a sampler is driven into the bottom of a boring by a jack or by a hammer. If there is any friction between the internal wall of the sample and the sample the action of driving in the assembly may have the effect of pushing back the material and the sample will turn out to be too short: hence the necessity for plenty of entry clearance. If the sampler is too thick, the action of driving in the assembly may have the effect of compressing the soil inwards and the sample will turn out to be too long: hence the use of thin walled samplers and the installation of a piston at the head of the sample whilst it is being lifted. In any case the extremities of the sample should be considered to be disturbed, and it is normal not to carry out tests on either end of a sample for a distance equal to one diameter. At the present time there is a great diversity of samplers perfectly adapted to different varieties of soils. Muds and loose cohesionless soils provide the most difficult conditions. In certain particularly important cases samples have been taken under a sludge placed in the borehole and attempts have been made to inject the soil with a polymerisable resin, or to freeze the soil, but the results have not always proved to be successful.

The size of test samples depends very much on the structure of the soils. Standard cylinders generally have a diameter of between 35 and 50 mm, but there is no reason why, in homogeneous soils, very much smaller cylinders, for example of 10 mm in diameter should not be used. Conversely, compressibility tests have been carried out in natural soils for which permeability was mainly dependent on structure on intact samples 250 mm in diameter. Indeed Marsal (1965) carried out triaxial tests for the El Infiernillo dam on stony material using cylinders 2 m in height. These cases illustrate the extreme limits for representative samples.

Faced with these difficulties, and that of conserving density, different *in situ* tests have been developed to determine the mechanical properties of the soil: penetration tests (driving a rod in the ground with separate measurements of the point bearing capacity and skin friction) (Sanglerat, 1972), vane test (shearing strength measured at the bottom of a hole by resistance to torsion), pressuremeter test (radial expansion of a boring) and lastly the Standard Penetration Test (Terzaghi and Peck, 1948) which consists simply of counting the number of blows from a drop weight required to drive in a sampler. This last test is particularly useful in pure sand where it is often difficult to take samples without modifying their density. In fact though, the chief advantages to be gained from *in situ* tests are those of economy which means that the tests can be repeated, and speed. Experience shows, however, that difficulties of interpretation may arise in numerous cases, especially if there are influences from nearby structures or if the strata are not very thick. In calibrating a set of measurements from one or more borings it would therefore be practically impossible to dispense with the taking of intact samples. After all, permeability tests apart, a knowledge of the fundamental properties of a soil cannot be gained by *in situ* testing alone. It must be accompanied by laboratory tests on intact samples run under perfectly known conditions.

3

INTRODUCTION TO THEORETICAL SOIL MECHANICS

Figure 3.1 is a typical example of the stress/strain curves resulting from triaxial tests on sand and clay. These curves are approximately linear at first, and then asymptotically approach the 'threshold of resistance'. It is this that justifies the use of the theories of elasticity and plasticity in soil mechanics.

3.1 Soil elasticity

The initial part of the stress/strain curve satisfies Hooke's law (proportionality of stresses and strains). Thus, around any one point Young's modulus and Poisson's ratio can be defined. In general, however, reversion does not take place since part of the soil deformation process is permanent. The use of elasticity theory in soil mechanics is therefore linked with the restrictive condition of monotonic stress. This restriction is of little importance in civil engineering practice and construction, but if it is to be transgressed (e.g. in the foundations of a silo which is to be alternately filled and emptied) further specialised study is needed.

Deformation of the points of contact between grains is not linear; the elastic modulus depends on the mean stress. In particular, in sands and normally consolidated clays the elastic modulus increases with depth. Thus, for a given surcharge, one must envisage an elastic solid in a certain volume whose behaviour is tangential to that of soil.

Fig. 3.1. Stress/strain curves illustrating the behaviour of sand and clay soils.

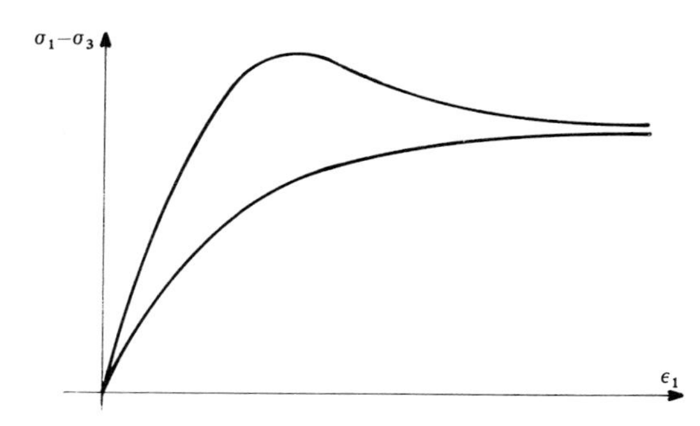

In this way it is possible to define the parameters E (Young's modulus) and ν (Poisson's ratio), which characterise the entire deformation of the soil mass. Furthermore the coefficients E' and ν' can be defined which characterise the elastic behaviour of the solid skeleton of the soil, in other words the granular structure. The parameters E and ν correspond to the instantaneous deformations, E' and ν' to long-term deformations when the pressure of the pore water, u, has once again returned to zero.

In a rapidly loaded saturated soil the volume, V, is practically constant. One can write:

$$\left.\begin{aligned} \epsilon_1 &= \frac{1}{E}\,[\sigma_1 - \nu(\sigma_2+\sigma_3)], \\ \frac{\Delta V}{V} &= \Sigma\epsilon_i = \frac{\sigma_1+\sigma_2+\sigma_3}{E}(1-2\nu) = 0, \end{aligned}\right\} \tag{3.1}$$

where ϵ_1 is the vertical component, and ϵ_2 and ϵ_3 are two orthogonal lateral components of strain; σ_1 is the vertical component and σ_2 and σ_3 are two orthogonal lateral components of stress; $\nu = 1/2$ because $\Sigma\sigma_i$ is not equal to zero. In the same way one can write for effective stresses ($\bar{\sigma} = \sigma - u$):

$$\left.\begin{aligned} \epsilon_1 &= \frac{1}{E'}\,[\bar{\sigma}_1 - \nu'(\bar{\sigma}_2+\bar{\sigma}_3)], \\ \frac{\Delta V}{V} &= \Sigma\epsilon_i = \frac{\bar{\sigma}_1+\bar{\sigma}_2+\bar{\sigma}_3}{E'}(1-2\nu') = 0, \end{aligned}\right\} \tag{3.2}$$

but this time ν' is not 1/2 since the solid skeleton is compressible. Thus

$$\Sigma\bar{\sigma}_i = \Sigma(\sigma_i - u) = \Sigma\sigma_i - 3u = 0,$$

from which, finally,

$$u = \tfrac{1}{3}(\sigma_1+\sigma_2+\sigma_3) = \sigma_{\mathrm{m}}.$$

This relation signifies that as long as there is no transfer of water, that is to say no consolidation, the pore pressure u is equal to the mean stress σ_{m}. Equating expressions (3.1) and (3.2) and reducing gives

$$\frac{1+\nu}{E} = \frac{1+\nu'}{E'},$$

or rather

$$\tfrac{2}{3}E = \frac{E'}{1+\nu'}.$$

This relation between the four parameters is very important theoretically. Any deviation from this expression should be interpreted as being the result of non-linear deformation (elastic or plastic). Once this relation is verified it leads to very convenient calculations. Let us take as an example the case of a rigid punch with a circular base applied to the surface of an elastic semi-infinite body. Using the coefficients E and ν, the theory of elasticity gives the initial penetration as

$$w_{\text{initial}} = \frac{P}{2RE}(1-\nu^2),$$

where P is the load and R is the radius of rigid base. The final displacement is obtained in the same way with the coefficients E' and ν':

$$w_{\text{final}} = \frac{P}{2RE'}(1-\nu'^2) = \frac{P}{2RE}(1+\nu)(1-\nu').$$

The settlement, i.e. the delayed displacement from the initial instant, is thus:

$$w_{\text{delayed}} = w_{\text{final}} - w_{\text{initial}} = \frac{P}{2RE}(1+\nu)(\nu-\nu') = \frac{3}{4}\frac{P}{RE}\left(\frac{1}{2}-\nu'\right).$$

3.2 Soil plasticity

The resistance threshold is not as easily detected in earth as it is in certain metals such as mild steel. In particular, when the stress/strain graph shows a maximum much greater than the values that follow (as in densely packed sands or structurally strong clays), interpretation becomes difficult and the phenomenon of progressive rupture may occur. However in the majority of cases an elasto-plastic model can be defined and more often than not the rigid–plastic model is sufficient.

The failure criterion is given by Coulomb's law

$$\sigma_t = c + \sigma_n \tan\varphi.$$

For a sand, then, we have simply (fig. 3.2).

$$\sigma_t = \sigma_n \tan\varphi,$$

or again

$$\frac{\sigma_3}{\sigma_1} = C = \frac{1-\sin\varphi}{1+\sin\varphi} = \tan^2(\tfrac{1}{4}\pi - \tfrac{1}{2}\varphi) = i = K_a$$

Fig. 3.2. Rupture criteria for a purely cohesionless soil ($c = 0$).

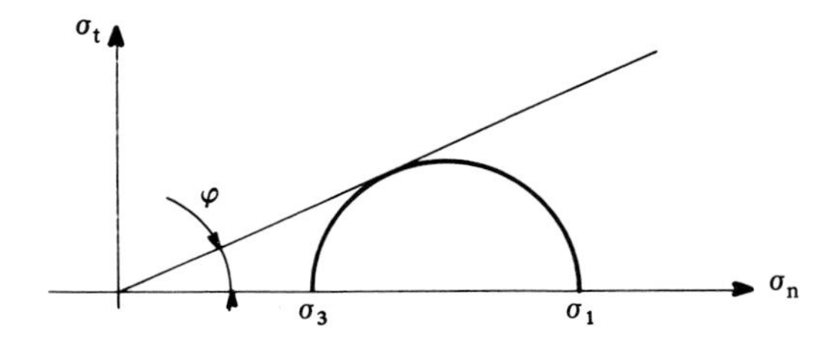

such that for

$$H = \frac{c}{\tan \varphi}$$

Coulomb's law in the most general case takes the form

$$\frac{\sigma_3 + H}{\sigma_1 + H} = i.$$

It can be seen that in this way the study of any soil can be reduced to that of a cohesionless material by translating the axis by H (fig. 3.3). This mathematical artifice is known as the principle of corresponding states, it reveals the same behaviour at failure of a cohesive mass, and of a cohesionless mass with the same angle of internal friction and subjected to an external hydrostatic pressure

$$H = \frac{c}{\tan \varphi}.$$

In the very important practical case where $\varphi = 0$ results can always be obtained by passing to the limit of solutions found for cohesionless mediums. The following exposition will therefore be limited to cohesionless mediums.

3.3 Semi-infinite mass, Rankine equilibrium, conjugate elements

Consider a semi-infinite earth mass inclined at ω. Forces acting on homologous points are equal (fig. 3.4). Writing the equilibrium of the forces and moments acting in the middle of the base AB of the small element shown in the figure gives the stress acting vertically on a face parallel to the free surface as

$$\sigma = \gamma z \cos \omega,$$

where z is the vertical distance of the element to the surface. When the mass is in ultimate equilibrium this relation is known as Rankine's hypothesis; it can only be strictly proved if the mass is assumed to be infinite. If the stress on a face parallel to the surface is vertical, then the stress on a vertical face is parallel

Fig. 3.3. Rupture criteria for a natural soil ($c \neq 0; \varphi \neq 0$).

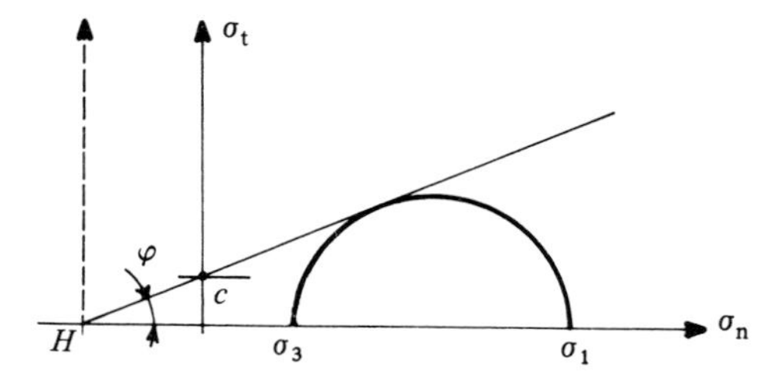

to the free surface and the two elements are conjugate (property of the ellipse governing the stresses).

The representation on Mohr's diagram is particularly interesting (fig. 3.5).

OE is the vertical stress acting on the face AB, therefore it is inclined at ω. The conjugate stress is OF' (symmetrical to OF) and is also inclined at ω. The direction of the major principal stress makes the angle $\frac{1}{2}(\hat{\gamma} - \omega)$ with AB (the angle $\hat{\gamma}$ is shown in fig. 3.5).

When $p = OC$ (mean pressure) then $CE = p \sin \varphi$ (radius of Mohr's circle) and the relation between the conjugate stresses can be written

$$OE = p \cos \omega - p \sin \varphi \cos \hat{\gamma},$$

$$OF = p \cos \omega + p \sin \varphi \cos \hat{\gamma},$$

$$\frac{OE}{OF} = i_\omega = \frac{\cos \omega - \sin \varphi \cos \hat{\gamma}}{\cos \omega + \sin \varphi \cos \hat{\gamma}},$$

Fig. 3.4. Rankine equilibrium.

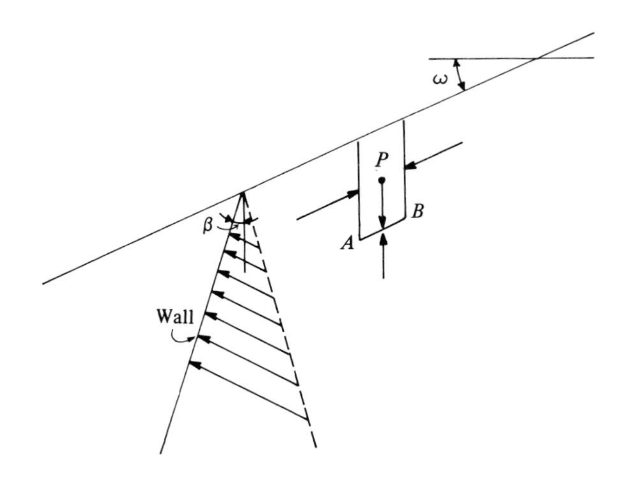

Fig. 3.5. Representation of Rankine equilibrium using Mohr's diagram.

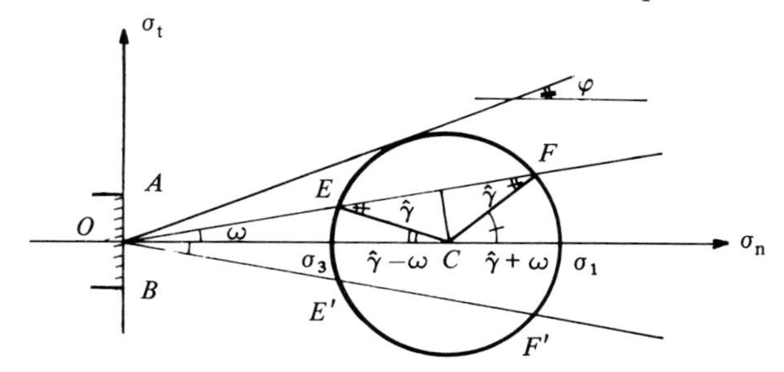

which with

$$\frac{\sin \hat{\gamma}}{p} = \frac{\sin \omega}{p \sin \varphi},$$

yields

$$i_\omega = \frac{\sin (\hat{\gamma} - \omega)}{\sin (\hat{\gamma} + \omega)}.$$

For $\omega = 0$ one finds again

$$i = \frac{1 - \sin \varphi}{1 + \sin \varphi}.$$

The value of the stresses on a face in any direction can be found by means of Mohr's circle in Rankine equilibrium. It is clear that at every point on a wall inclined at an angle β (fig. 3.4) to the vertical the stresses are parallel and proportional to the distance from the intersection with the surface.

3.4 Upper and lower equilibrium

In reality the determination of the conjugate stress is not completed by using the preceding arguments since there are in fact two circles passing through point E tangential to the intrinsic curve (fig. 3.6). Thus, in Rankine's solution, two states of limiting equilibrium exist for a mass of earth. These are the upper and lower equilibrium or active earth pressure and passive earth pressure – they correspond to the following physical phenomenon.

Imagine a frictionless, vertical wall in a horizontal mass of sand (fig. 3.7).

Fig. 3.6. Upper and lower equilibrium.

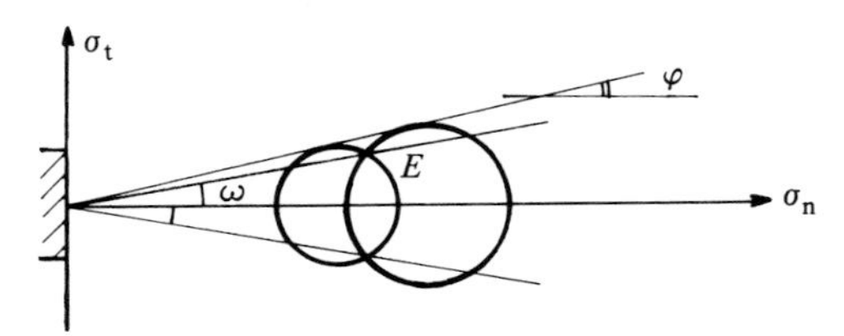

Fig. 3.7. Active thrust and passive resistance.

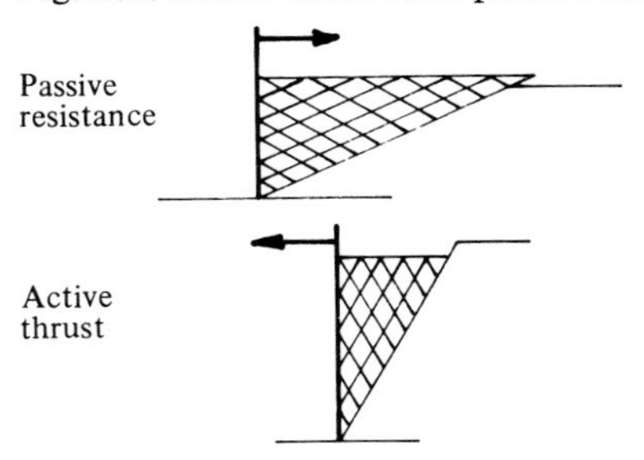

The wall is subjected to a stress (indeterminate) called the earth pressure at rest. Eliminate the left part of the mass and move the wall horizontally to the right. It produces in the sand a plastic deformation and is now in ultimate equilibrium, which is the passive Rankine state. The stress now acting on the wall is called passive resistance (passive earth pressure). Move the wall to the left. The mass is now in the active Rankine state and the stress on the wall is the active thrust (active earth pressure). In both cases the slip lines are straight.

3.5 General case

In order to respect Rankine's theory a frictionless, rigid vertical wall has been introduced into a horizontal mass. In the most general case the wall is neither vertical nor frictionless and the free surface of the mass is not horizontal. As there is a relative displacement of the medium and the wall, the directions of the stresses are determined by the wall and not by Rankine's solution. The mass cannot be considered to be infinite, Rankine's theory no longer applies and the slip lines are not straight. In the active Rankine state they in fact differ little, but in the passive Rankine state they are so far from being straight that Rankine's solution cannot even be considered as an acceptable approximation.

The general case, connecting Rankine equilibrium at a distance from the wall with the equilibrium in the vicinity of the wall, has been studied by Boussinesq. He retained the hypothesis of stresses proportional to the radius vector (which comes down to saying that the resultant of the forces is applied to the lower third of the wall), and he presented a series of differential equations (non-integrable) for the equilibrium of a heavy wedge of soil. Résal completed the approximate integrations for Boussinesq's equations for several cases of the active Rankine state. Finally Caquot developed Boussinesq's method and came up with precise calculations for active and passive Rankine states (Caquot and Kérisel, 1966). The results of these calculations are presented in charts or tables where the active and passive states are given as a function of:

- ω slope of the free surface;
- β inclination of the wall;
- α angle the resultant makes with the wall;
- φ angle of internal friction.

It can be seen that when $\beta = \frac{1}{2}\pi$, $\omega = 0$ the way is opened to the theory of the bearing capacity of foundations.

The charts in fig. 3.8 permit the active thrust and passive resistance behind a retaining wall to be determined in the very important case where the free surface is horizontal ($\omega = 0$) both for masses with and without a surcharge and for cohesive and cohesionless soils. When calculating a real case where the soil has a water table, the submerged density, γ_i, should be used for the submerged soil, and of course, the hydrostatic thrust on the internal face of the wall should be added.

Fig. 3.8. Active and passive earth pressure charts (after Caquot and Kérisel, 1948) for the case $\omega = 0$.

Passive resistance at the angle $-\varphi$ | Active thrust at the angle φ

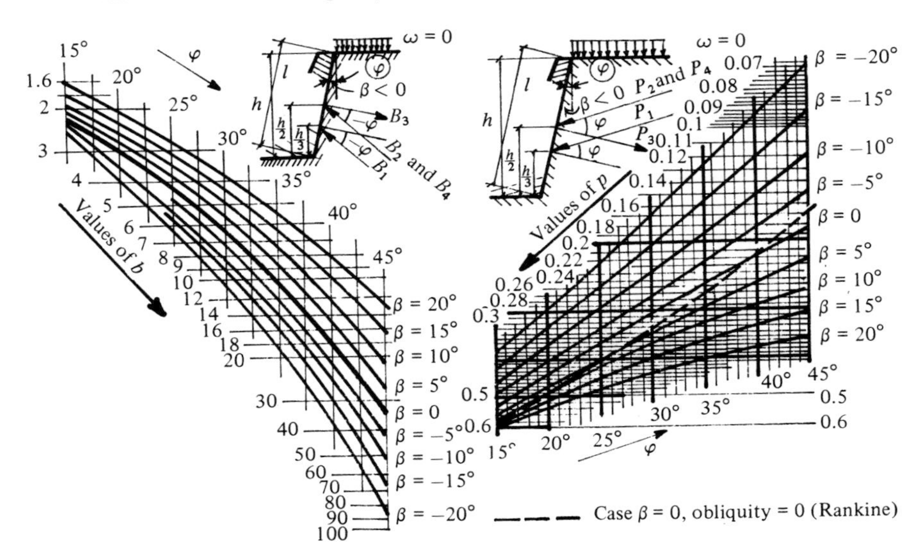

Passive earth pressure on a purely cohesionless mass:

$$B_1 = \tfrac{1}{2}\gamma l^2 b$$

(b is given by the graph).

Increase in passive earth pressure due to cohesion:

$$B_1 + B_3.$$

Increase in passive earth pressure due to a normal surcharge:

$$B_4$$

$$B_3 = -Hl = -cl \cot\phi$$

$$B_2 \text{ and } B_4 = q \tan\left(\tfrac{1}{4}\pi + \tfrac{1}{2}\phi\right) \times e^{2\delta \tan\phi}$$

whence

$$\delta = \tfrac{1}{4}\pi + \tfrac{1}{2}\phi - \beta$$

$$q = H = c\cot\phi \text{ for } B_2$$

$$q = \text{surcharge for } B_4.$$

Active earth pressure of a purely cohesionless mass:

$$P_1 = \tfrac{1}{2}\gamma l^2 p$$

(p is given by the graph).

Decrease in active earth pressure due to cohesion:

$$P_2 + P_3.$$

Increase in active earth pressure due to a normal surcharge:

$$P_4$$

$$P_3 = -Hl = -cl \cot\phi$$

$$P_2 \text{ and } P_4 = q \tan\left(\tfrac{1}{4}\pi - \tfrac{1}{2}\phi\right) \times e^{-2\delta \tan\phi}$$

whence

$$\delta = \tfrac{1}{4}\pi - \tfrac{1}{2}\phi - \beta$$

$$q = H = c\cot\phi \text{ for } P_2$$

$$q = \text{surcharge for } P_4.$$

3.6 Similarity in soil mechanics

Models are so often used in association with elasticity (bridges, dams, vibrations, etc.) that it is not necessary to emphasise the advantages of using similarity methods in soil mechanics. Their role is even more important in the area of plasticity and rupture. It is possible to resolve plasticity problems in two dimensions by resorting to long and onerous numerical calculations (two differential equations for equilibrium and the equation of failure criteria are needed to determine two principal stresses and their direction, thus three unknowns). The theory is found wanting in three dimensions: here there are too many unknowns for the number of equations (i.e. three general equations for equilibrium, and the equation of failure criteria to determine six unknowns for the stresses); the use of a small-scale model is thus of particular value. Several techniques are possible for studying the quasi-static rupture of earth. The similarity relations that emerge from the general equations of equilibrium and the behavioural relationships must be observed. Let $\overset{*}{u}$ denote the scale of a quantity, that is $u_{\text{model}}/u_{\text{structure}}$ (Mandel, 1966), For the equilibrium equation

$$\Sigma_j \frac{\partial \sigma_{ij}}{\partial x_i} + \rho g_i = 0$$

(where ρ represents the density) to hold when the scales are modified, it is necessary that

$$\overset{*}{\sigma} \overset{*}{l}^{-1} = \overset{*}{\rho} \overset{*}{g}.$$

As it is difficult to modify the densities ($\overset{*}{\rho} = 1$) and since the scale for length is given, it can be seen that two possibilities exist:

(*a*) In the first case the gravitational acceleration is conserved ($\overset{*}{g} = 1$) and we find

$$\overset{*}{\sigma} = \overset{*}{l}.$$

In small-scale models approaching rupture the deformations are large; in that case the scale for length should be equal to the scale for displacement in order to respect the geometrical changes; this results in the scale for deformations

$$\overset{*}{\epsilon} = 1.$$

The conditions for rheological similarity can easily be found using the equations for elasticity or plasticity:

elasticity: $\overset{*}{E} = \overset{*}{\sigma}, \overset{*}{\nu} = 1;$

plasticity: $\overset{*}{c} = \overset{*}{\sigma}, \overset{*}{\varphi} = 1$ (similarity of intrinsic curves).

These equations define a new material, called an equivalent material, which is easy to fabricate if the phenomenon studied is completely elastic or completely plastic. This becomes more difficult if both the deformations and rupture have to be taken into account since the four preceding relations must then be verified simultaneously. If the elasticity is not linear (flattening of the stress/strain curve between the linear portion and the plastic threshold, variations of volume before failure, etc.) the similarity becomes more and more approximate. In the end, if the properties change from brittle (rupture by fraction and formation of fissures) to ductile (plastic flow) when the mean pressure increases, it is generally impossible to find a convenient material.

(*b*) In the second case the material itself is used to construct the model, with the same stresses ($\overset{*}{\sigma} = 1$). The difficulties encountered in the preceding paragraph no longer arise and one can be certain that the rheological relations are observed. The equilibrium equations then give

$$\overset{*}{g} = \overset{*}{l}^{-1}.$$

The forces acting on a volume of the mass must then be increased, by centrifuge, for example, or by the hydraulic action of seepage forces (Zelikson, 1967), or by inserting cables, sealed at points within the mass and then stretched (Oberti, 1957). However, one must ensure that there is no scale effect on the mechanical properties; in fissured materials, for example, such as rocks or particular types of clays, the smaller the sample the greater the resistance. The same is true for dense cohesionless media where the variation of the thickness of the slip surface (dilatancy) is a function of the size of the grains and not the scale of the model, etc.

In the particular case of pure, dry sands, the intrinsic curve passes through the origin; the $\overset{*}{\sigma} = \overset{*}{c}$ condition in paragraph (*a*) disappears: similarity can thus be achieved while conserving both the material and gravity. But the stresses are not the same and the similarity for displacements is not correct since the elastic modulus of cohesionless materials is a function of the mean stress. Further, as the critical density depends on the stress, the position of γ in relation to γ_c is modified, which alters the comparison of the volume variations. Such a similarity cannot, therefore, be accepted except where the rigid–plastic model is acceptable. Finally, more refined models can be used that take into account dynamic effects, viscosity, water flow in the soil, etc. In general, however, perfect similarity is not entirely possible, and certain conditions have to be neglected. One might be led in the extreme to full-scale tests; this eventuality is discussed in chapter 9 in relation to the bearing capacity of piles.

4

EQUILIBRIUM OF SLOPES AND RETAINING WALLS

4.1 Calculating the stability of a structure

More often than not the application of the theory of plasticity leads to difficult mathematical constructs, and from these to long numerical calculations that are difficult to transfer to a computer. The theoretical solutions that are available at present are limited either to geometrically simple problems (plastic equilibrium around a hole) or to problems encountered on such a frequent basis that their resolution is practically indispensable (active and passive earth pressure, foundations, etc.) or to problems of little practical importance (stable, curvilinear slopes etc.).

As a result of this engineers have turned to methods of approximation whose use is sufficiently simple to be adapted to practical everyday problems, but whose solutions are sufficiently precise not to differ greatly from rigorous solutions. In this way, by making assumptions about the formation of a wedge of earth behind an overthrown retaining wall, of circular slip lines in a slope, and of a slip line in the form of a logarithmic spiral under a punch, Coulomb, Fellenius and Rendulic have all put forward simple methods. It is of course essential to know the order of magnitude of the error associated with adoption of these assumptions. The following theoretical considerations provide satisfactory justification.

A field of stress $\sigma(x, y, z)$ is said to be statically admissible if, at all points, it satisfies the conditions for equilibrium. It is said to be plastically admissible if at all points it respects the conditions for rupture. A field of deformation velocities $v(x, y, z)$ is said to be kinematically admissible if it is compatible with the continuity of the body and with external connections. It is said to be plastically admissible if its divergence is zero.

The moment ultimate equilibrium is crossed, the field of stress is plastically and statically admissible and the field of velocities is kinematically and plastically admissible.

In the case of a purely cohesive body it can be rigorously proved that the load corresponding to ultimate equilibrium, called the limit load, is equal to the largest external load that gives a field of statically and plastically admissible stresses, and to the smallest external load that corresponds to a kinematically and plastically admissible field of deformation velocities.

In practice it is rare to find either the largest or the smallest of these loads, and the skill of the designer lies in finding the closest upper and lower bounds. Different examples are examined below using the hypothesis of a soil of cohesion c and with an angle of internal friction $\varphi = 0$ (clay rapidly subjected to loading).

Example 1: Indentation at the surface of an undefined mass

The theoretical solution has been found by Prandtl; the ultimate pressure q_u is given by $q_u = (2 + \pi)c$.

(*a*) *Static method.* Fig. 4.1 shows a particularly simple way to cut the continuous media under the indenter. To the right of Y (and to the left of Y') the equilibrium is assured by Mohr's circle of simple compression, and between Y and Y' by the Mohr's circle that is tangential to this. The field of stress corresponding to $p_1 = 4c$ is statically and plastically admissible. In the present case one can be certain that $p_1 < q_u$ since the field of velocities is not kinematically admissible: along the length of Y a segment elongates or becomes

Fig. 4.1. Finding a lower bound for the limit load of a punch acting on the surface of a purely cohesive mass.

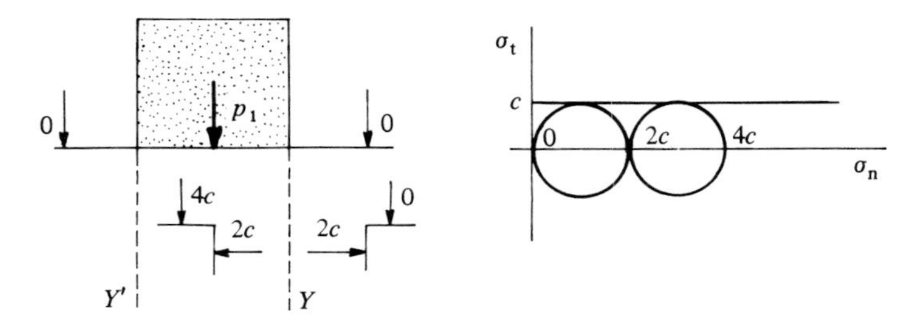

Fig. 4.2. Finding a lower bound close to the limit load of a punch acting on the surface of a purely cohesive mass.

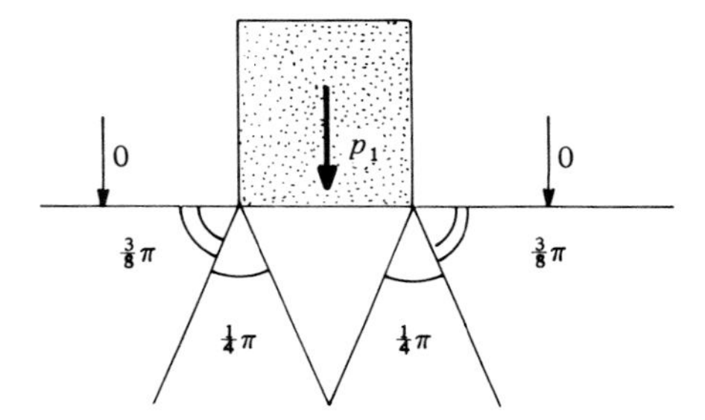

shorter depending on whether it is situated on the right or left of the medium: thus continuity is not assured.

The method of cutting the soil under the indenter illustrated in fig. 4.2 is not much more complicated and gives in the same way $p_1 = 4.83c$.

(*b*) *Kinematic method.* Imagine a displacement of the entire body about a point O (fig. 4.3). The dissipated energy is localised in the slip circle. The field of velocities in this plastic flow is kinematically and plastically admissible. Writing the equilibrium of the moments acting at O:

$$p_2 \cdot B \cdot (B/2) = c \cdot 2R\alpha \cdot R,$$

$$p_2 = 4c\,R^2\alpha/B^2 = 4c\,\alpha/\sin^2\alpha.$$

All that remains is to calculate the value which minimises $\alpha/\sin^2\alpha$ (that is $\alpha = 67°$), to find $p_2 = 5.56c$, and thus it is proved that $p_1 < q_u < p_2$.

Example 2: Simple compression

(*a*) *Static method.* It is evident that there is a statically and plastically admissible field of stress: this is the uniform field with $\sigma_x = \sigma_y = 0$ and $\sigma_z = 2c$ (fig. 4.4*a*).

(*b*) *Kinematic method.* To have a field of deformation velocities that is kinematically admissible let the upper section shown in fig. 4.4*b* slide in a plane

Fig. 4.3. Finding an upper bound for the limit load of a punch acting on the surface of a purely cohesive mass.

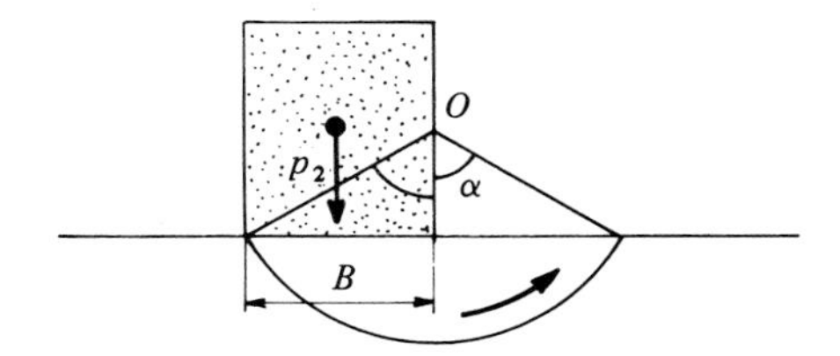

Fig. 4.4. Simple compression: (*a*) static field; (*b*) kinematic field.

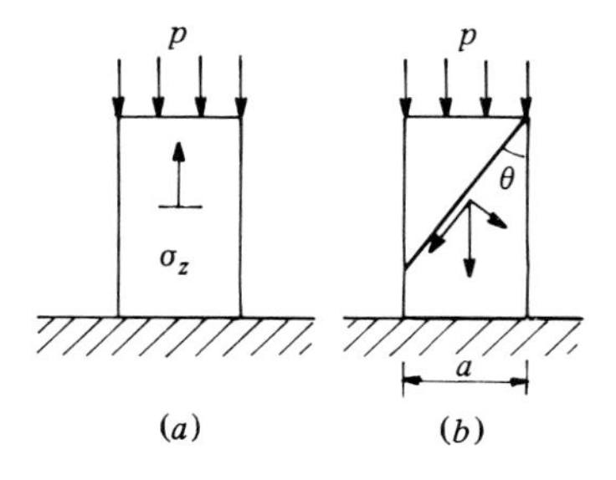

whose path is shown by the straight line inclined at θ to the vertical. By projection on this line we have:

$$pa \cos \theta = ac/\sin \theta ,$$

$$p = 2c/\sin 2\theta ,$$

which is a minimum when $\sin 2\theta$ is a maximum. Let $\theta = \frac{1}{4}\pi$; with this slip plane, $p = 2c$. There is no distinction between the static and kinematic solution; it is thus the exact solution.

Example 3: Equilibrium of an underground structure near to the surface

The underground structure is a circular hole of radius R whose centre is situated at a depth H. The soil is purely cohesive and loaded at the surface by a pressure p. Our examination will be restricted here to the kinematic method corresponding to a vertical slip towards the base of the division bounded by two parallel planes indicated in fig. 4.5. The half-angle in the centre, θ, is taken as a parameter. By projection on the vertical axis we find

$$pR \sin \theta = c(H - R \cos \theta),$$

whence

$$p = c\left(\frac{H}{R} - \cos \theta\right)\left(\frac{1}{\sin \theta}\right).$$

In order to minimise p it is necessary to cancel the derivative p'_θ:

$$p'_\theta = c\left[\sin^2\theta - \left(\frac{H}{R} - \cos \theta\right) \cos \theta\right]\left(\frac{1}{\sin^2\theta}\right)$$

$$= c\left(1 - \frac{H}{R} \cos \theta\right)\left(\frac{1}{\sin^2\theta}\right) = 0,$$

Fig. 4.5. Tunnel in a purely cohesive medium (kinematic method).

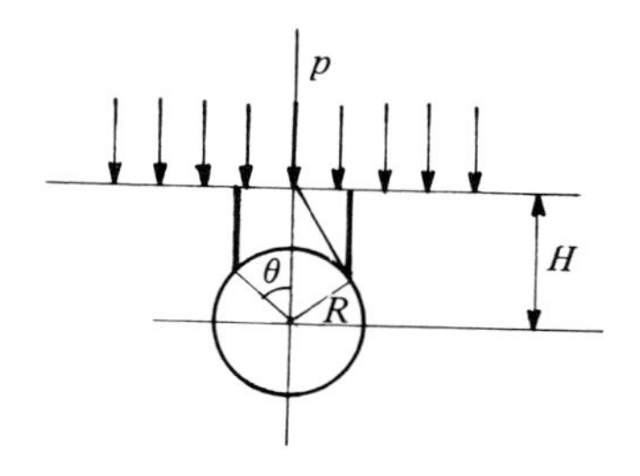

so that $\cos\theta = R/H$. The simple geometric construction in fig. 4.5 gives the least favourable plane; the limit load becomes:

$$p = c\left(\frac{H}{R} - \frac{R}{H}\right)\left(1 - \frac{R^2}{H^2}\right)^{-1/2}$$

$$= \frac{c}{R}(H^2 - R^2)^{1/2}$$

$$= c\left(\frac{H^2}{R^2} - 1\right)^{1/2}.$$

In this way the kinematic method provides an easily found, simple result. This formula is acceptable for excavations near to the surface. Other models of plastic flow are needed for excavations at greater depth.

4.2 Calculating the stability of a slope

It would seem that the closer the field of deformation velocities is to the real field, the closer the load p_2 is to the limit load p. Calculating the stability of a structure or a slope is thus a trial and error procedure, seeking out the least favourable slip line. This slip line is almost always taken to be circular and the best known method is that of Fellenius.

The principle is remarkably simple. A circular slice is cut out of a certain number of vertical elements whose weight is in equilibrium under the action of the lateral forces (which cancel each other out in pairs) and by the reaction along the slip line (fig. 4.6). If the weight of the element is resolved into the components N_i and T_i along the slip line, it is seen that N_i does not affect the equilibrium of the moments about the centre O of the circle. The relation between the motive and the resisting moments is thus:

$$F_s = \frac{\tan\varphi\Sigma N_i + cl}{\Sigma T_i},$$

where l is the length of the slip line. In this way a safety factor F_s can be defined in relation to the resisting moment. The least favourable circle, that is the circle

Fig. 4.6. Slip circle method.

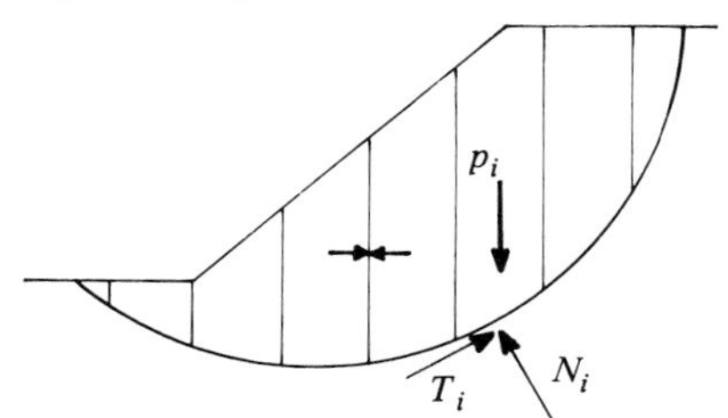

whose safety factor is the smallest, can be found by a series of successive tests. The concept of safety is here fairly complex. Other definitions may be used such as safety in relation to φ and c, or in relation to the maximum height of a stable slope. Whatever its use, it must be remembered that the safety factor is determined by a kinematic method and as such has no rigorous physical meaning. In particular, one can easily see that in a cohesionless slope inclined at φ and of finite height, thus in the immediate neighbourhood of ultimate equilibrium (that is when the safety factor is close to 1), a safety factor of 1.05 to 1.10 (depending on the value of φ and on the geometry of the slope height) is obtained when Fellenius' method is applied.

Calculation by the method of slices is a bit lengthy and it is often advisable to use a planimeter. However, this method is particularly useful for heterogeneous slopes or for slopes in the presence of water. In fact, the calculation is unchanged if one uses total stresses and uplift pressures. It has been established that since the latter are normal to the slip circle they do not have a motive moment. Thus, this method is extremely convenient when studying the equilibrium of slopes under the effects of seepage networks or in the particularly unfavourable condition of rapid drawdown.

If in homogeneous masses one considers a circle of friction of radius $R \sin \theta$ the calculation can be simplified, but in this case the tables of Fellenius, or Terzaghi, or Taylor immediately give the least favourable circle and the safety factor. Finally it should be noted that the slip circle method lends itself easily to calculation by computer.

4.3 Retaining walls

Calculations for retaining walls are made by determining the active earth pressure behind the wall. In the initial state the force behind the wall is the earth pressure at rest, which depends on the soil packing and is generally unknown. Under the action of this force the wall is displaced, and the earth pressure is diminished until it reaches a limiting value given by a Rankine state – thus failure is reached. However, since the failure takes place on a slip line that is

Fig. 4.7. Calculating Coulomb's wedge.

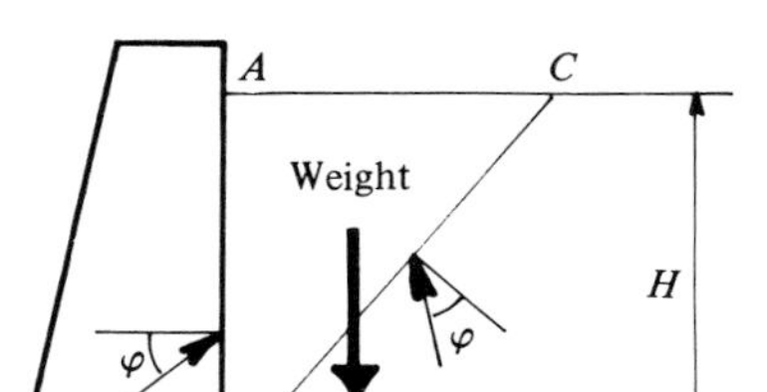

almost straight, Coulomb's method is in most cases completely acceptable (fig. 4.7).

The wedge *ABC* under the action of gravity is in equilibrium with the frictional force along *BC* and the reaction of the wall, which is at an angle φ. It suffices to find the inclination of *BC* that gives the largest reaction, and one arrives at:

$$P = \tfrac{1}{2}\gamma H^2 \frac{\cos\varphi}{(1+\sqrt{2}\sin\varphi)^2}.$$

The case of the inclined wall and a mass with a sloping free surface has been studied by Poncelet using the same hypotheses. With the notation and symbols of fig. 4.8, the active pressure is given by:

$$P = \tfrac{1}{2}\gamma l^2 \frac{\cos^2(\beta-\varphi)}{\cos(\beta+\alpha)}\left[1+\left(\frac{\sin(\varphi+\alpha)\sin(\varphi-\omega)}{\cos(\beta+\alpha)\cos(\beta-\omega)}\right)^{1/2}\right]^{-2}.$$

Finally, a retaining wall transmits a stress to its foundation. The verification of general stability with regard to deep sliding must not be forgotten. The only possible method of calculation here is once again the slip circle (fig. 4.9).

Fig. 4.8. Poncelet's calculation (the sign convention adopted is such that in this figure the angles α, β, ω are positive.

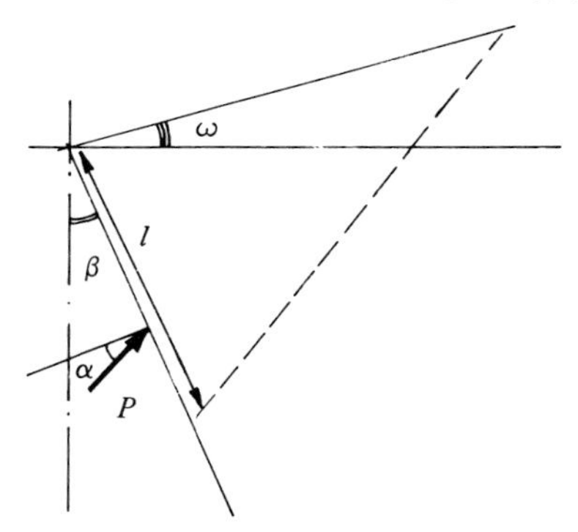

Fig. 4.9. Checking the stability of a retaining wall and its foundation.

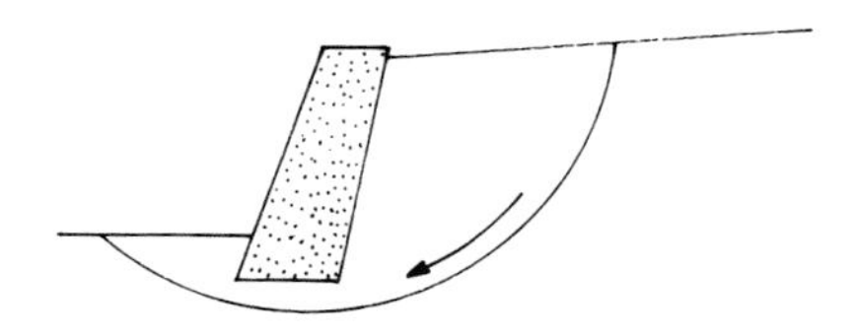

4.4 Landslide stabilisation

The practical calculation of the stability of natural slopes and of retaining walls is in reality more delicate than the theory would lead us to believe. In fact, information is often lacking on the mechanical characteristics and hydrodynamic regime of the water table. Also, for natural slopes one does not generally have sufficient material means to study them.

On the contrary, when the terrain does slip, the situation is completely changed. Not only can one carry out a geotechnical study, but above all one has at one's disposal information from a truly large-scale experiment, from which, using acceptable hypotheses, one can write an equilibrium equation where the safety factor is equal to unity. Consequently, the stabilisation operations possible can be estimated at their exact value by incorporating them into the failure equilibrium equation and seeing the improvement brought about in the safety factor.

Often a change in the flow regime of the water table causes an accident in the terrain; so when a natural slide occurs it is necessary to undertake a complete examination of the mode of flow of infiltrations.

There are only a few procedures available for the stabilisation of landslides, amongst which are the following, listed in order of importance:

- drainage:
  - shallow trenches following the steepest line
  - deep drainage
  - long term – planting of vegetation (notably acacias)
- earth transportations:
  - loading the base
  - unloading the top
- retaining walls
- anchorages

Water often plays a major role in the collapse of walls (Logeais, 1982) and landslides. Figs. 4.10 to 4.17 show this influence in widely differing examples.

Fig. 4.10. There is no small retaining wall (the weephole has been left out).

Fig. 4.11. Despite the snow overburden the wall is holding; only just, but it is holding.

Fig. 4.12. After the snows have melted the wall has been flattened (the drainage outlet has probably frozen).

Fig. 4.13. A permeable crib holds perfectly.

Fig. 4.14. Following persistent rain the side of an embankment slope has collapsed taking with it a benching 1.5 m wide and a piece of the pavement.

Fig. 4.15. Mechanism of regressive erosion of the bank of a road cutting away the natural slope.

Fig. 4.16. The sloping trees indicate an area of active subsidence.

Fig. 4.17. Landslide turning into a flow of mud (liquefaction). Vertical height: about 800 m.

5

ROADS AND BRIDGES

5.1 Road alignment

Economic considerations are not the only constraining factors in designing the course of a road running between two specific points: the design is also constrained by the relief of the terrain, the curvature to be allowed, slope limits, etc. Slopes may be lessened by constructing embankments. In the past, engineers tried to achieve a balance between excavations and embankments since dumping or opening a quarry were onerous operations. Nowadays, with modern equipment, it often makes better sense to reduce the transport distance even at the cost of neglecting the balance between excavations and embankments. Whatever the situation, however, transport of earth must be kept to a minimum.

Once the course of the road passing through the necessary points has been defined then visible areas of inferior soil, low lying areas and wet areas are determined in order to anticipate drainage necessary to remove water from the pavement: hydrologic examination (stream, spring flow, position of the water table in winter and summer etc.) is thus necessary.

This is followed by an examination of the soils along the course of the road to define the borrow zones to be used as a source of construction material for the pavement, following criteria which will be defined below. The final plan for the road takes into account the amount of traffic, the load of the vehicles and extra constraints that may or may not be acceptable (partial submersion in the case of flooding, thaw barriers etc.).

In general, in a varied region it is usually possible to find materials possessing all the necessary qualities to build foundations and filter beds.

5.2 Drainage

Water is capable of destroying the majority of structures. It tends to lessen the mechanical resistance of soils and thus of pavements. In addition, the presence of water brings with it the risk of freezing in winter. The pavement must therefore be protected from water. Suitable drainage, with ditches and outlets allowing the gravitational flow of rain water, is of the utmost importance. Places where the water table runs level with the soil surface need special treatment with gravitational drains. If this is impossible an embankment should be constructed to keep the inferior soil away from the pavement. The road is then superelevated without ditches and with a lateral flow of rain water. In such a case there must be a stabilised watertight foundation.

5.3 The embankment and its foundations

Embankments built for keeping inferior soil separate from the pavement or negotiating accidents of relief can be either compacted or not compacted.

If the natural soil is inferior (soft clay, organic soil, etc.) care must be taken not to build the bank to such a height that there is a failure in the foundation. Soft clays can often be assimilated with purely cohesive bodies and the limit load is $p = (2 + \pi)c$: should the load of the embankment be greater than this, loading berms can be placed laterally or gentle slopes can be used so that the embankment is superelevated. In working with such soils one must take into account settlement under the foundations in the years following construction and make provision for this possibility. Whenever possible peat should be eliminated, for example by punching with the weight of an embankment above the projected structure. In certain cases the settlement can be accelerated by using vertical sand drains (a hexagonal centered network with spacing 1.5 to 2 m) and temporary loading embankments.

Generally speaking, the construction of a low embankment does not lead to any foundation problems, but when the soil is not compacted the settlement of the body of the embankment itself must then be taken into consideration. An unsaturated soil used in an embankment progressively settles under its own weight. One can easily find out the order of magnitude of the settlement by an oedometric test with a saturated sample. On the other hand, there is at present no sure method of determining the amount of settlement in a particular period of time. An embankment of average size will often take at least six months to stabilise (that is in effect a season of rain); this process can take a significantly longer time if the embankment is high and contains a certain amount of clay. It is obviously advantageous then, to construct the embankment as early as possible before carrying out further work.

5.4 Bridges and other road constructions

The problem of the bearing capacity of foundations will be considered elsewhere. The discussion here will be restricted to generalities and since retaining walls have already been considered, we will limit ourselves to roadway bridging structures.

When the planned road encounters a small water course the simplest expedient is to lay pipes (often metallic) of a suitable diameter running through the embankment that are flexible enough to follow the movements of the soil and the embankment. The advantage of this technique lies in avoiding too precise a study of the soil for a structure that would indeed be small.

In crossing a larger stream, one of the better ways of avoiding the inferior soils of low lying areas is to construct the bridge a small distance from the stream, and thereafter divert the stream. Diverting the water course can both improve the flow as well as simplifying the laying of the foundations. The original bed of the stream can then easily be filled in.

Finally, for a larger structure where a bridge is accessed from an embankment, it is always advisable to build the embankment before the bridge itself. The weight of the bridge is much less than that of the embankment, and there is always the risk of differential settlement between the bridge and the embankment that might lead to problems later on.

5.5 The pavement

One can simplify the road structure by distinguishing three superimposed layers: the soil (natural or embankment), the base, and the running surface. With the exception of soft mud and peats, the soil has to be taken very much as it is found. The soil level is sometimes referred to as the subgrade. The base layer is sometimes divided into the base and sub-base. The role of the sub-base is simply to avoid placing an expensive foundation of high bearing capacity on mediocre soil. The base is a good quality layer with low deformability under load such that the running surface can work in the best conditions and remain in good shape. An example of what may happen with a poorly constructed base is shown in fig. 5.1. One of the more important functions of the base and sub-base is to drain the roadway so that it is protected from moisture and frost.

Fig. 5.1. Foundation failure. Remedy: lift the coating and all the base layer as far as the subgrade and construct a new pavement.

Fig. 5.2. Three examples of pavements carrying heavy traffic.

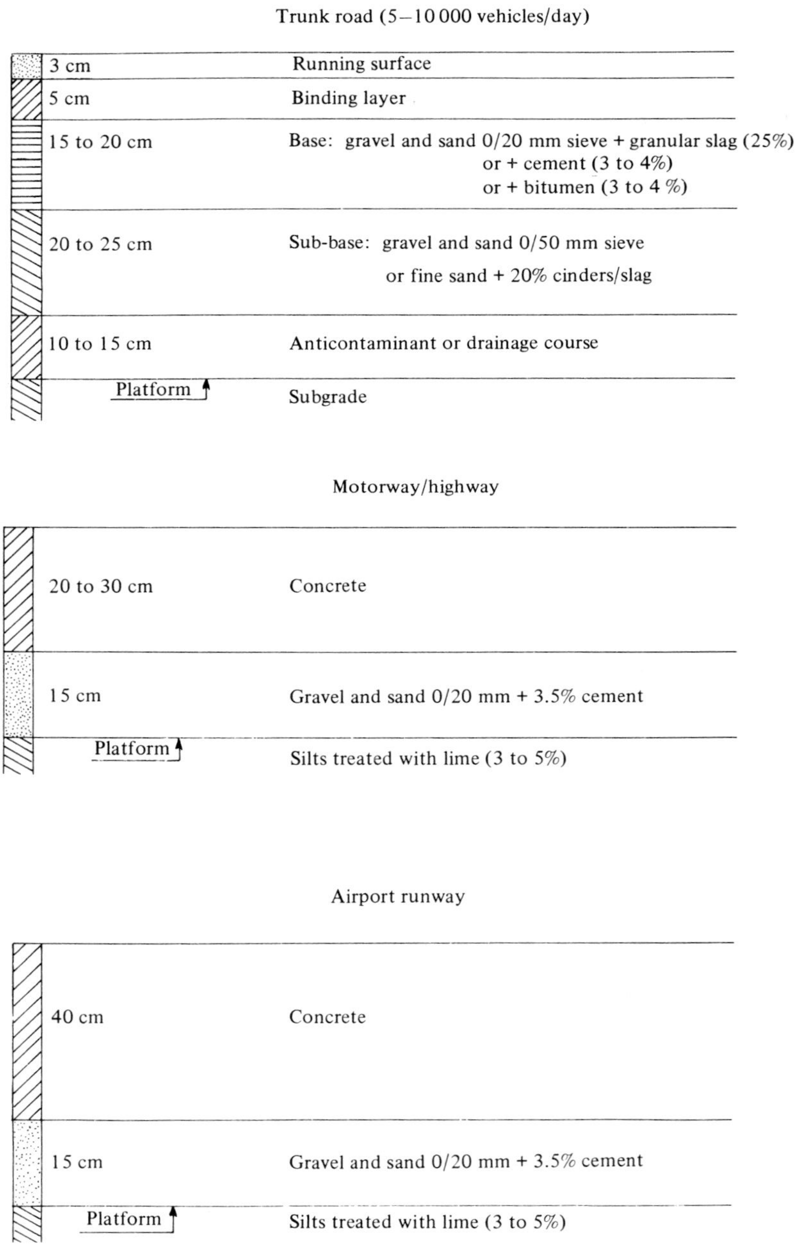

Trunk road (5–10 000 vehicles/day)

3 cm	Running surface
5 cm	Binding layer
15 to 20 cm	Base: gravel and sand 0/20 mm sieve + granular slag (25%) or + cement (3 to 4%) or + bitumen (3 to 4 %)
20 to 25 cm	Sub-base: gravel and sand 0/50 mm sieve or fine sand + 20% cinders/slag
10 to 15 cm	Anticontaminant or drainage course
Platform	Subgrade

Motorway/highway

20 to 30 cm	Concrete
15 cm	Gravel and sand 0/20 mm + 3.5% cement
Platform	Silts treated with lime (3 to 5%)

Airport runway

40 cm	Concrete
15 cm	Gravel and sand 0/20 mm + 3.5% cement
Platform	Silts treated with lime (3 to 5%)

The running surface protects the base from rain water. It is watertight and can be constructed from bituminous concrete, 'black top' materials, or from a stabilised soil. If constructed from the former, there should be a 'tack coat' and a 'wearing coat' whose surface should be non-skid even in wet weather.

Rigid running surfaces such as those made from concrete used for heavy traffic (heavy both in number and weight) and aerodromes have not been discussed here since the technique is radically different from that used for flexible pavements (see fig. 5.2).

5.6 Design of a pavement

The rational calculation of flexible pavements is not yet sufficiently developed to work out the design from non-empirical methods alone. This is due in part to the fact that road failure is not simply a problem of plasticity, but also illustrates the phenomenon of fatigue associated with the fabrication of fines (fillerisation). In these conditions the area where linear behaviour ceases is ill-defined, and the value of the mechanical coefficients defining the elastic behaviour of the soil and conditions for rupture are hardly known at all.

Numerous empirical methods have been proposed; their value and scope depends on climatic zones, and transposing a method from one climate or soil to another without precautions can lead to catastrophe. The Californian bearing ratio method (CBR) first applied by the American army during the Second World War is one of the best known and most widely used methods.

The CBR test consists of punching at a constant rate to gain an indication of the resistance of the material. A sample of the material is placed in a cylindrical mould 15.2 cm (6 in.) in diameter and 15.2 cm deep at the density and water content envisaged *in situ*. A circular load is placed around the punch to simulate the weight of overlying layers. The test has been strictly standardised in the USA. The punch is cylindrical. Its cross-sectional area is 19.3 cm^2 (3 sq. in) and it is pressed into the soil at a rate of 1.27 mm/mn ($\frac{1}{20}$ in. per minute). The stress/penetration diagram is drawn with, in some cases, a correction for the initial deformation.

The index of 100 is given as a frame of reference for a very good material which, under a punching pressure of 6.9 MPa sinks by 2.54 mm ($\frac{1}{10}$ in.) and under 10.32 MPa sinks 5.08 mm ($\frac{2}{10}$ in.). The CBR index can be worked out for the material studied as the percentage of the index 100 punching pressure required to give the same penetration.

The CBR method uses graphs drawn up from many years of experience which show the thickness required for different wheel loads (fig. 5.3).

If a, b, c, are the CBR indices for the soil, sub-base and base, the graphs give the thicknesses A, B, C, (fig. 5.4):

A is the total thickness of the pavement;
$A - B$ is the thickness of the sub-base;
$B - C$ is the thickness of the base;
C is the thickness of the running surface.

The determination of the thickness of the pavement by the CBR index is an empirical and approximate method. It does not define the water content at which the test should be carried out. Saturating the soil in water for four days prior to the test is certainly pessimistic in many cases. It would be better to arrange for the soil to have the equilibrium water content it would have after a certain time as governed by *in situ* conditions, i.e. by its distance from the water table and possible capillary rise.

The CBR method neither takes into account the traffic, nor does it give any indication of the nature of soils to use. These are specified by the following recommendations, themselves empirical.

The materials used for the base should neither be friable nor susceptible to frost; they should neither contain organic matter nor clods of clay. Their liquid limit should be less than 25% and their plasticity index less than 6; it is thus a question of using pit-run materials containing a predominance of cohesionless materials, gravels or sands. However when the PI is less than 4 its nature is very difficult to determine experimentally: below this level the plasticity index is practically impossible to measure. One then uses a sand equivalent (SE) to distinguish between sands. This is determined by a sedimentation test – the fraction of sand that passes through a 0.5 mm sieve is put into a test tube containing a shaken flocculant solution for twenty minutes. A deposit of sand

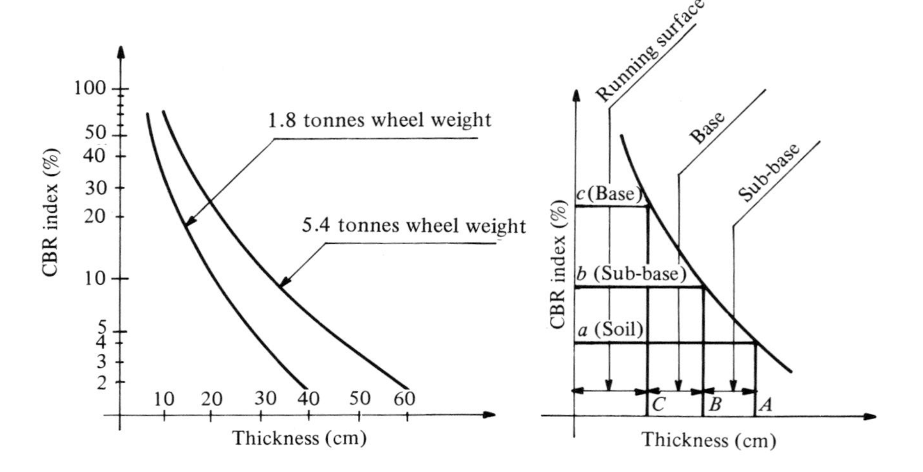

Fig. 5.3. CBR graphs.

Fig. 5.4. Working out the dimensions of a pavement by the CBR method.

of height h_2 forms in the floculated material of height h_1 above which lies clear water. By definition SE = $100\,h_2/h_1$. If SE > 30 the material is not plastic. If SE < 20 it is clearly clayey. In general, if one has the choice between several materials for the base of the pavement, one should choose that which has the greater sand equivalent value and one should also try to use material with a grain size distribution that has been well tried in other road constructions locally. Futhermore, the percentage of elements passing through the AFNOR 20 sieve (0.08 mm) should be less than $\frac{2}{3}$ of the percentage passing through the AFNOR sieve (0.4 mm).

If a hydrocarbon coating is not used, the running surface should be a little more clayey in line with the following limits: the percentage of material passing through the AFNOR 20 sieve should be greater than 8%. The liquid limit should be less than 35%. The plasticity index should lie between 4 and 9.

In the light of the above rules the following comments can be made: pebbles of some 50 mm can easily be tolerated in the sub-base. Gravel–sand–clay mixtures are used in the base and the cohesion and plasticity of the running surface should be increased as a function of the climate. In fact, a gravel–sand–clay mixture cannot endure the alternation of wetting (loss of resistance, swelling) and drying out (dust, fissuration). If the climate is very dry it is advantageous to increase the liquid limit (w_L) to get a higher cohesion, even if we must accept putting a smaller quantity of clay in the gravel–sand–clay mixture so that there is some permeability. In a damp climate however, the plasticity is lowered (w_L), while making sure, of course, that the medium is still permeable. It is of course obvious that a pavement without a running surface will not tolerate rain and traffic: it is only worth economising on the running surface when constructing a temporary road or track.

It is more difficult to specify precautions to be taken against frost. Soil freezing is a complex phenomenon determined by numerous parameters: the intensity of the cold, the duration of exposure to frost, the depth of the water table, the porosity of the soil, and also the quantity of rain falling just before the frost. It manifests itself by swelling of the soil, an increase in the solid water content and by the pumping of water vapour from the water table (cold wall principle), and sometimes by the formation of ice lenses. Obviously the road can be used when the soil is frozen, but when the thaw sets in disasters can occur if the road is used before the water has drained away naturally (thaw barriers). The choice of an appropriate grain size distribution adapted to the climate is thus not a complete guarantee of success. Casagrande has recommended that continuous graded soil (at Hazen's coefficient $D_{60}/D_{10} > 5$) should have less than 3% by weight of elements finer than 20 μm, and uniform grain size soil ($D_{60}/D_{10} < 5$) should have less than 10% by weight of elements finer than 20 μm.

This ruling has been proposed as standard in the United States but should not be accepted categorically in France. The best precaution available is still to overlay the original material below frost level with a material not susceptible to frost, and perhaps separate the two by a blanket of clean sand to avoid contamination; the maximum depth of freezing in France is of the order of 60 cm. It can be seen that in this way many instances arise where the total thickness of layers exceeds that recommended by the CBR method.

6

COMPACTION

6.1 Principle of compaction

Whatever the method used for packing the earth–passage of machines, rollers, compaction by static loading, mechanical or manual tamping, the final density of a soil depends on its water content. If the procedure remains rigorously the same, then one can identify a maximum dry density ($\gamma_{d\max}$) at an optimum water content (w_{opt}) on the compaction curve. To the left of this value on the curve shown in fig. 6.1 there is not enough water to ensure that the grains have enough lubrication to settle properly in place. To the right of the optimum there is too much water for the grains to pack in the closest contact with each other. If the compaction energy is increased the curves converge; $\gamma_{d\max}$ increases and w_{opt} decreases. The plane (γ_d, w) is bounded on the right by the hyperbola defined by the parametric equation

$$w_s = \gamma_w n/\gamma_s(1-n),$$

$$\gamma_d = \gamma_s(1-n).$$

called the saturation curve; here n is the porosity and γ_s the specific gravity, as defined in sections 1.3.4 and 1.3.5.

Compaction is studied in the laboratory using the standard compaction test known as the Proctor test. The method used for compaction is completely arbitrary; it has moreover been modified several times, so that there are different

Fig. 6.1. Compaction curves.

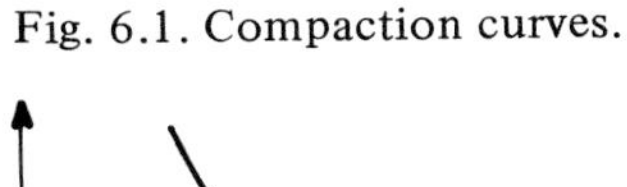

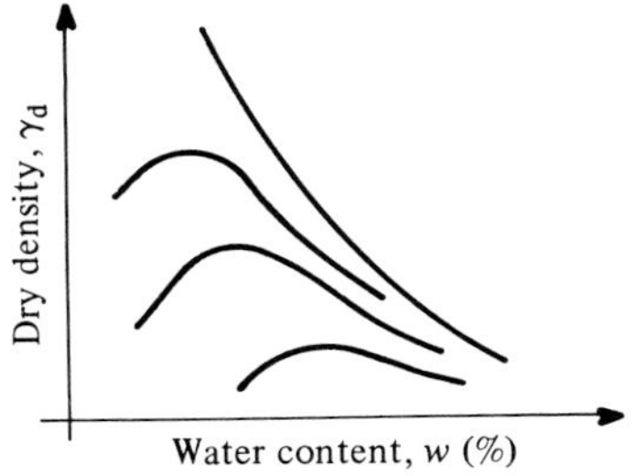

standards. The two most important are shown in the table below. The soil in a metallic mould or so-called 'Proctor mould' (diameter 10.2 cm (4 in.), height 11.7 cm, (4.6 in.); volume 0.96 dm^3) is compacted by a blow from a ram. In both cases the soil is struck off level with the rim in the mould before weighing and the test repeated with varying water contents. The compaction energy is about five times greater in the modified Proctor test than in the standard Proctor test; it is perhaps ten times smaller than used in on-site tests which are executed in an entirely different manner, and thus difficult to compare with the Proctor tests described here.

The tests are conducted in the laboratory on particles that are smaller than 5 mm which makes the comparison with reality even more tenuous. In any case the correspondence between the density attained using machinery in the field and that of the standard Proctor test in the laboratory should be confirmed by an *in situ* test whose results can subsequently be used when checking site work.

Generally speaking, the equipment used for road compaction gives dry densities of the order of 95% of the maximum dry density of the standard Proctor test, or 90% of the maximum dry density of the modified Proctor test. Heavy wheeled rollers of the type used for compacting earth dams will give 95% of the maximum dry density of the modified Proctor test and sometimes even more.

Trying to achieve the maximum dry density is similar to trying to find the best resistance of the compacted soil. It should be noted that from a practical point of view if two materials are equivalent, the one which gives a flatter maximum on the curve will allow a greater margin of error in the water content than that which has a sharp peak. The exercise should thus be carefully supervised and preference given to the former.

For the purposes of soil identification, if a modified Proctor test yields a maximum dry density value below 1.6 the soil is often unsatisfactory, a value between 1.8 and 1.9 is satisfactory, and a value of 2.05 or more is excellent. In the same way, an optimum water content greater than 20% is an unfavourable

	Normal Proctor	Modified Proctor
Mass of ram	2.49 kg	4.53 kg
Height of fall	30.5 cm	45.7 cm
Thickness of layers	4 cm	2.5 cm
Number of layers	3	5
Number of blows per layer	25	25

sign. But such rules cannot have a universal truth. What is important is the mechanical characteristics of the soils compacted to a certain density, which we shall now examine.

6.2 Physical properties of compacted soils

6.2.1 *Resistance*

The resistance of a compacted soil to simple compression or to penetration by a needle is a decreasing function of the water content.[1] It would seem that there is an advantage in being on the left of the curve at constant γ_d. In fact this is not true since after saturation (which is always likely to occur) and in the absence of any swelling (fig. 6.2) the state of the two soils A and A' evolves to point B and the same final product is obtained no matter what their origin.

If the soil is subjected to a surcharge, with no external source of water, it settles. The representative point A on fig. 6.3 passes through C, then to D with the expulsion of water. If this expulsion is difficult, consolidation no longer takes place and the shearing resistance stops increasing.

If the final equilibrium is at D, then it is sufficient to start from an initial water content less than that at D and take, for example, the state at E. By reasoning of this type one can regulate the water content during the construction of high, compacted embankments to avoid the development of dangerous pore pressures.

6.2.2 *Permeability*

The volume of voids and thus the permeability of the soil is reduced by compaction. This is easily shown in laboratory tests. Compaction can, in certain cases reduce the permeability by a factor of ten.

Fig. 6.2. Compaction at the same dry density.

Fig. 6.3. Evolutions of a compacted soil.

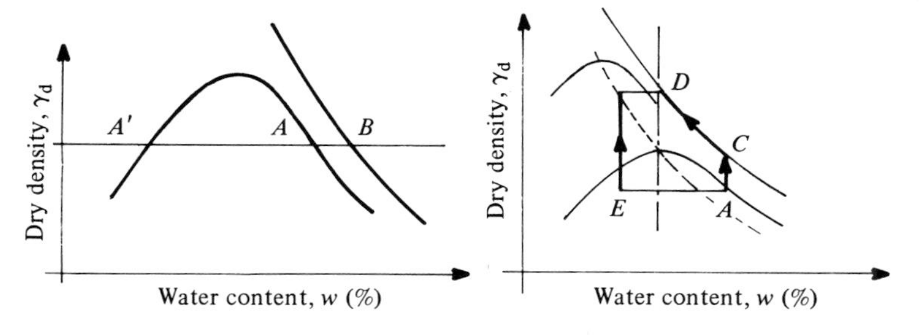

1 The CBR test is a standardised punching test. The CBR index depends on the degree of compaction.

6.3 Compacting equipment

The following types of equipment are used for compacting in the field:

smooth wheel roller;

sheepsfoot rollers (whose mass of 15 to 20 tonnes is limited by the excessive power needed to haul or drive them);

rubber-tyred rollers (mass 50 to 200 tonnes – inflation pressure 0.4 to 0.7 MPa);

vibrating plates (from 0.5 to 5 tonnes – frequency 10 to 20 Hz – area of plate from 0.5 to 1 m^2);

smooth vibrating rollers;

percussion machine and self-propelling rammers such as the frog-rammer.

This equipment is designed to harden the soil by passing over it a certain number of times, which for economic reasons should not be too many, but on the other hand should not be so few that the bonding between neighbouring compacted zones is not satisfactory. The heavier the roller then the larger the surface contact with the soil, the effect will be felt at greater depths and the compacted layers can be thicker.

Smooth rollers or rubber-tyred rollers are most suitable for heavy clays; vibrating plates are not at all advisable since they do not have a very deep effect, although the surface appears excellent after they have been used. When the water content of a clayey soil is too high lamination sometimes occurs as the number of passes increases. This phenomenon, incorrectly termed overcompaction, is harmful and one should break up layers where this has happened. Sandy materials can be compacted by smooth-wheeled rollers, vibrating rollers or vibrating plates. It is advisable to use spraying and vibration simultaneously when compacting sands. The presence of water conducts the vibration far and wide and the water gives a cohesion which permits the compaction of the cohesionless surface layer. Lastly, self-propelling rammers give excellent results in the majority of cases; however they are not used very often in France.

The important factors to be taken into account and monitored in carrying out the operation are, essentially, the number of times the equipment passes over the zone, the water content at the site of compaction and lastly the dry density attained.

6.4 Application to roads

In road construction the aim is to obtain a high CBR index. This can be ensured in different ways.

6.4.1 *Mechanical stabilisation*

It is compaction above all that improves the bearing capacity of the soil, the sub-base and base. If no natural terrain with acceptable mechanical properties can be found, different types of soils can be mixed together to obtain a particle size distribution typical of 'geoconcrete', a concrete made of gravel, sand and earth. The result is satisfactory only if the equipment ensures an intimate close blend.

6.4.2 *Stabilisation by adding cement*

Adding 8 to 10% cement by weight to the soil results in a lean concrete whose resistance to simple compression can reach 1 to 2 MPa. The major difficulty with this is reaching a sufficiently blended mix on the site – a difficulty which almost completely eliminates the possibility of treating heavy clays in this manner. Less cement is added to gravels and sands (3 to 4%) than to other soils. The soil must be well-compacted since this has a major influence on the resistance of the final product.

The optimum quantity of cement to be added to the soil can be worked out in the laboratory by examining the strength at intervals of 7 and 28 days of different Proctor moulds. This is generally carried out at optimum water content, w_{opt}, while making certain that there is no organic matter present that may prevent the cement from setting. The durability of the resultant mixture can be estimated by a standard test of energetic brushing.

Other materials are also able to bring about a chemical hardening comparable to that of cement: raw slag, fly ash and lime.

6.4.3 *Chemical stabilisation*

Numerous chemical products have been used to stabilise soils: bitumen (hot or cold), organic resins (Vinsol, Rosine, etc.) lignosulphates, acrylates, amines, aniline/furfural mixture, water repellants, silicones, etc.

The results vary from one soil to another. Apart from bitumen they are expensive products used in very small quantities. Each time they are used research has to be undertaken into the optimum dosage. This research can easily be done in the laboratory using the methods already mentioned. The percentage of bitumen used is of the order of 3 to 4%. If bituminous emulsions are used this can be increased to 6%.

6.5 Application to earth dams

In homogeneous earth dams, in dams with different earth zones, or in rockfill dams with a watertight core of earth, a preliminary study of the soil's capacity for compaction is indispensable.

6.5.1 *Watertight earth structures*

Imperviousness is an important quality for an earth structure. The soil best suited for this can be identified in the laboratory, although it is not necessary to use the most impervious. The dry density required *in situ* can likewise be determined for the soil finally chosen.

6.5.2 *Stable earth structures*

The mechanical qualities of a soil are improved by compaction. Resistance tests (direct shear, triaxial tests, etc.) measure this improvement to give the mechanical characteristics needed for calculating the inclination of slopes – by the slip circle method for example. Compressibility tests allow one to forecast the settlement of the structure under its own weight. If pore pressures can develop in stable structures then precautions must be taken, one of the more important ones being to ensure compaction of the soil to a lower water content than the optimum to avoid the risk of saturation. This method is however disadvantageous in that it leads to somewhat rigid embankments with a tendency to develop fissures. That is why in some cases, such as that of low embankments, it can be better to work at a water content higher than w_{opt}: if the soil is very clayey one thus avoids the danger of swelling; the soil is also more plastic which can be useful for example in constructing a watertight core; it is, in addition, more impermeable. Lastly, if, with this second method, there is any danger of pore pressure developing, one can always place pressure gauges in the embankment to monitor the development of pore pressure with loading and consequently one can adapt the timing of the construction of the work to dissipate the pore water pressure.

7

BEARING CAPACITY OF SHALLOW FOUNDATIONS

When a foundation is under an ever increasing load, at some point, plastic rupture takes place in the soil: failure occurs. The allowable bearing pressure is calculated by determining the limit load, or failure load, and then lowering this value by the factor F_s called the safety factor which will be discussed in more detail later. The calculation of the bearing capacity of foundations is thus based on the failure load.

It is necessary to distinguish between shallow foundations, the object of this chapter, and deep foundations – piles, shafts and caissons. We will use the following arbitrary definition: a foundation of breadth B, sunk to a depth D can be considered as shallow when $D/B < 5$ or 6.

The problem of indentation of a real soil by a foundation is an extremely complex one to which, except in very simple cases, there are no exact mathematical solutions. However, relatively precise approximations exist for present day practical foundation problems.

The difficulty of the problem necessitates considerable simplification, and we shall first of all make the following restrictive hypotheses, lifting them in due course:

the soil is an infinite half-space, homogeneous and isotropic;
the material is rigid–plastic (characterised by φ, c, γ);
the problem is two dimensional (loaded strip);
the applied force is central and vertical;
the foundation is not deformable.

7.1 Purely cohesive soils

The case of purely cohesive soils ($\varphi = 0$) is very important in current practice since it encompasses the rapid loading of all saturated plastic clays.

7.1.1 *Foundations on the surface of a purely cohesive soil* ($\varphi = 0$)

If a pressure is applied to the surface of a semi-infinite elastic body one can define the limit load as that which produces a shear stress equal to the resistance of the material. Thus, $q = \pi c$. If the body is brittle, this value corresponds to the failure load. If the body is plastic the stresses are redistributed and the load can be increased. Plasticity theory gives the same value for limit load for both rough-based foundations (Prandtl) and smooth-based foundations (Hill) (fig. 7.1), thus:

$$q_u = (2+\pi)c.$$

The allowable pressure is

$$q_a = (2+\pi)c/F_s$$

(F_s is the safety factor).

7.1.2 *Foundation sunk in a cohesive soil*

Above the level of the foundation the soil acts as a surcharge equal to γD (fig. 7.2). Consequently the failure load is simply increased by γD. Thus,

$$q_u = \gamma D + (2+\pi)c.$$

Similarly for the allowable pressure

$$q_a = \gamma D + (2+\pi)c/F_s.$$

Obviously the safety factor calculation should not be based upon the weight of the earth. It is easy to imagine a situation where such a mistake would lead to failure, but in the opposite direction. In particular, failure at the bottom of an excavation of depth D can occur (see fig. 7.2) for

$$\gamma D = (2+\pi)c.$$

It is assumed in the preceding equations that the only effect of the soil is its dead weight. In fact, shearing appears in the soil lying above the level of the

Fig. 7.1. Foundation at the surface ($\varphi = 0$): (*a*) Rough-based foundation (Prandtl); (*b*) smooth-based foundation (Hill).

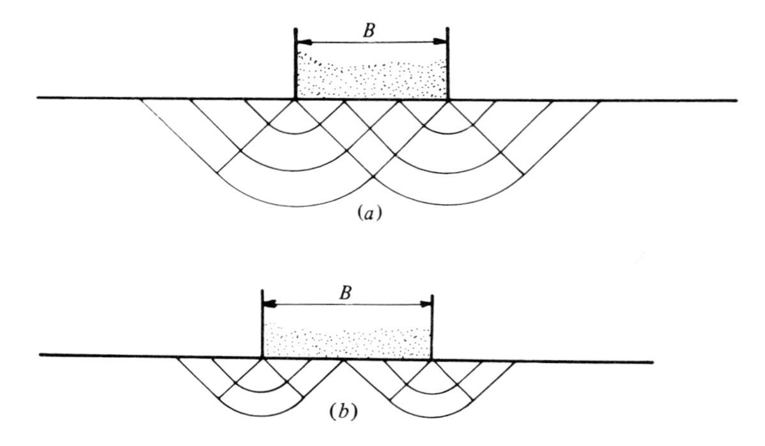

foundation. In these conditions, Skempton has proposed the use of a factor that increases linearly with depth, equal to 5.14 when $D/B = 0$ and 7.5 when $D/B \geqslant 5$.

7.1.3 *Isolated foundation on cohesive soil* ($\varphi = 0$)

The preceding solutions were valid for footings of indefinite length, that means in practice rectangular strip footings whose slenderness ratio (length/breadth) is greater than 5. The problem is three dimensional for isolated circular footings. In this case experience has shown that the limit load is higher and that the preceding result should be increased by a factor of 1.25. The difference in failure loads between square and circular footings is imperceptible if the areas of contact are equal. Finally, for sunken foundations the factor proposed by Skempton should be taken into account. Thus the failure pressure on the surface is

$$q_u = 6.4c,$$

which increases with depth to

$$q_u = 9.4c,$$

where $D/B \geqslant 5$.

Fig. 7.2. Sunken foundation ($\varphi = 0$) and critical depth D of an excavation.

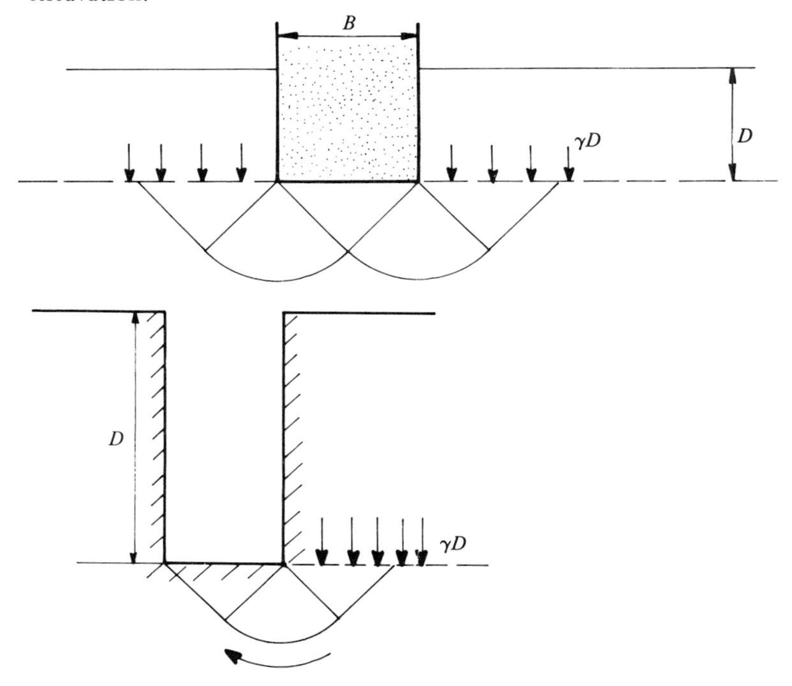

7.1.4 *Foundation resting on a cohesive bed of limited thickness* ($\varphi = 0$)

The clay layer is itself resting on a more resistant material that is assumed to be undeformable. If the load does not have its own tensile strength (surcharge of an embankment for instance) consideration of a circular failure line tangential to the base of the clay layer, emerging at 45° and centred on the edge of the load (kinematic method) leads to $q_u = 2\pi c$ (fig. 7.3*a*).

For a triangular embankment, the same calculation gives $q = 3\pi c$ which defines a limit slope since all the slip circles tangential to the base are equally unstable (Habib and Suklje, 1954) (figs. 7.3*b* and 7.3*c*). If the foundation does have resistance (rigid and rough bases) the presence of a shallow resistant layer

Fig. 7.3. Loading a bed ($\varphi = 0$) of limited thickness: (*a*) uniform load; (*b*) load increasing linearly from the edge (slope); (*c*) profiles of the increase of a colliery slag heap constructed on a cohesive bed of limited thickness.

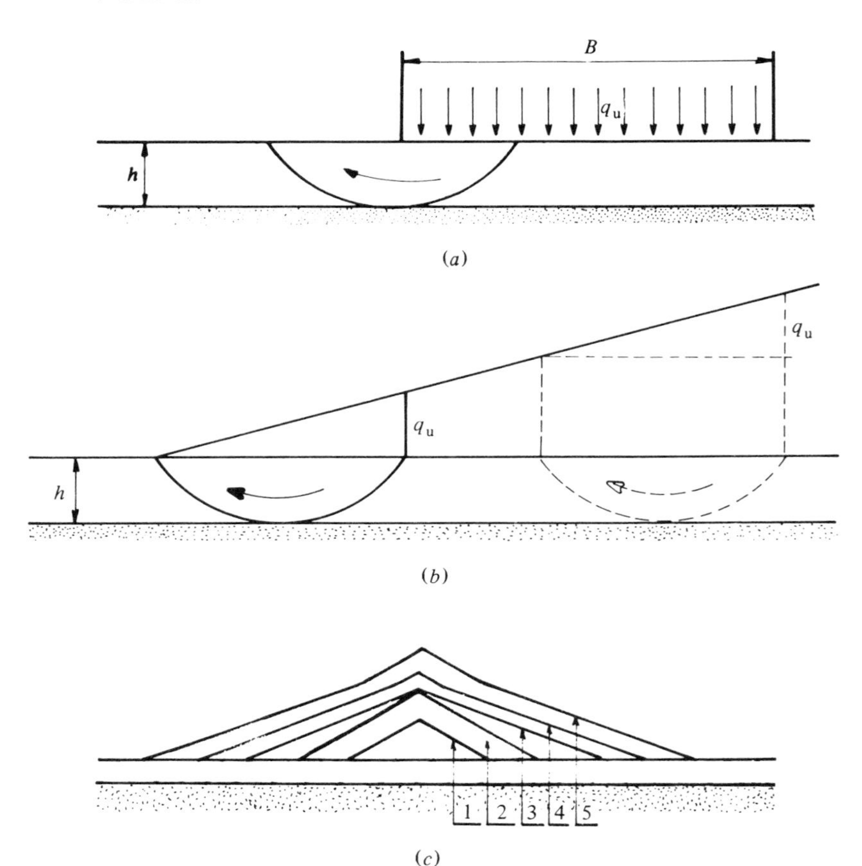

of thickness h considerably improves the bearing capacity and

$$q_u = c\,[(B/2h) + \pi + 1]$$

when $B/2h > 3$. The larger B/h the better the approximation (Mandel and Salençon, 1969).

7.2 Purely cohesionless soils (sands and gravels)

7.2.1 *Foundation at the surface of a cohesionless medium* ($c = 0$)

Prandtl's solution has been extended to bodies which have internal friction; the plastic flow network has a different form. The network illustrated in fig. 7.4 has two Rankine states connected by logarithmic spirals (rather than circles) and Prandtl's fan, and is a limiting case for a weightless, cohesionless mass. In a real cohesionless mass the net of slip lines is slightly different. The calculation of the exact network is difficult but has been done by Lundgren and Mortensen. As far as the ultimate pressure is concerned the important result is the fact that the breadth of the foundation intervenes linearly, which is quite obvious if one remembers that in Rankine equilibrium the stresses are proportional to the radius vector. The ultimate pressure is given by

$$q_u = \tfrac{1}{2}\gamma B N_\gamma,$$

where N_γ is a function of φ. This bearing capacity coefficient can be called the surface factor.

Fig. 7.4. Foundation at the surface ($\varphi = 0$). (*a*) Representation of a two-dimensional plastic flow and (*b*) corresponding theoretical model.

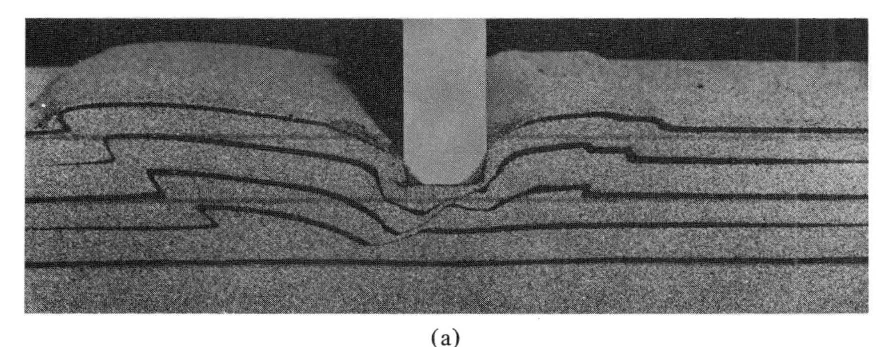

(a)

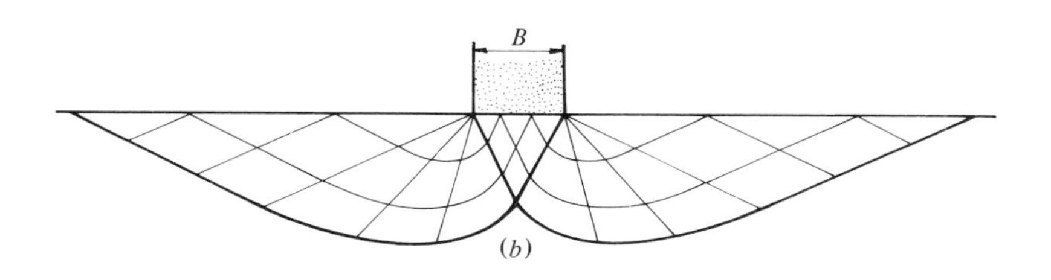

(*b*)

7.2.2 *Foundation sunken in a cohesionless soil* ($c = 0$)

In addition to the surface factor a supplementary resistance develops along the slip line, due to the presence of the surcharge γD (fig. 7.5). The corresponding bearing capacity is proportional to γD; it is written $\gamma D N_q$ where N_q is a function of φ and can be called the depth factor. The ultimate pressure is given by:

$$q_u = \tfrac{1}{2}\gamma B N_\gamma + \gamma D N_q.$$

In the network in fig. 7.5 the same simplification (i.e. that of weightlessness) has been made as in fig. 7.4. Salençon (1965) gives the rigorous calculation.

7.2.3 *Isolated foundation on cohesionless soil* ($c = 0$)

The preceding solutions were valid for footings of infinite length, that is, in practice, strip footings whose slenderness ratio is greater than 5. The problem is three dimensional for an isolated footing, and for a square foundation the limit load must be reduced by a factor that experience shows to be of the order of 0.7 for the surface factor. The shape of the footing does not seem to effect the depth factor, at least not at small depths.

Finally, as in the case of cohesive masses, one cannot distinguish between the failure loads for a square of circular foundation when the same area of surface is in contact with the soil.

7.3 Soils having cohesion and an angle of internal friction (natural soils)

7.3.1 *Foundation at the surface of a natural soil* (φ, c, γ)

In addition to the surface factor there is a supplementary resistance due to cohesion along the slip line (fig. 7.4). The corresponding term is proportional to c; it can be written cN_c where N_c is a function of φ called the cohesion factor. The ultimate pressure is

$$q_u = \tfrac{1}{2}\gamma B N_\gamma + cN_c.$$

7.3.2 *Foundation sunk in a natural soil* (φ, c, γ)

The general equation which gives the ultimate pressure under a

Fig. 7.5. Sunken foundation ($\varphi \neq 0$).

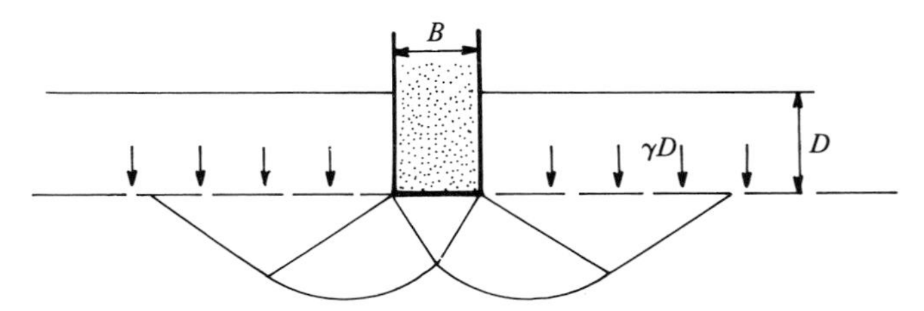

foundation on natural soil (two-dimensional problem, fig. 7.5) is

$$q_u = \tfrac{1}{2}\gamma B N_\gamma + \gamma D N_q + c N_c,$$

the values for N_γ, N_q, N_c are indicated in table 7.1 and on the chart in fig. 7.6.

Comments

(1) The factors N_γ, N_q, N_c given in table 7.1 have been obtained by semi-empirical methods by Terzaghi and other authors. It should be noted however that they can be theoretically justified in the framework of the theories of earth pressure. In particular, a reasonable approximation for the factor N_γ can be obtained by calculating the passive earth pressure along a surface inclined at $\frac{1}{4}\pi - \frac{1}{2}\varphi$ to the vertical at the edge of the foundation, a case which corresponds to the phenomenon, which experience shows to be frequent, of a wedge of earth in redundant equilibrium under a rigid foundation.

The factor N_γ corresponds then to the passive earth pressure inclined at an angle of φ to a wall such that $\beta = \frac{1}{4}\pi - \frac{1}{2}\varphi$ and with $\omega = 0$. The factor N_q can also be easily explained. It corresponds to the effect of the earth above the level of the foundation. Caquot has shown that one must have

$$q_u = \gamma D j e^{\pi \tan\varphi},$$

from which

$$N_q = j e^{\pi \tan\varphi},$$

where

$$j = \tan^2(\tfrac{1}{4}\pi + \tfrac{1}{2}\varphi) = K_p = 1/K_a.$$

As for the factor N_c, this can easily be obtained by using the theory of

Table 7.1. *Values for* N_γ, N_q *and* N_c.

$\varphi°$	N_γ	N_q	N_c
0	0	1	5.14
5	–	2	7
10	1	3	9
15	2	5	13
20	5	8	18
25	10	13	25
30	20	23	37
35	44	44	60
40	110	86	100
45	330	170	170

corresponding states. If a uniform stress equal to $H = c/\tan\varphi$ acts around a mass there is a simple surcharge on the level of the foundation and one can write

$$H + q_u = Hj\mathrm{e}^{\pi\tan\varphi},$$
$$q_u = H(j\mathrm{e}^{\pi\tan\varphi} - 1),$$

whence

$$N_c \; \frac{j\mathrm{e}^{\pi\tan\varphi} - 1}{\tan\varphi} = \frac{N_q - 1}{\tan\varphi}.$$

Fig. 7.6. Factors of the general equation giving the ultimate pressure $q_u = \frac{1}{2}\gamma BN_\gamma + \gamma DN_q + cN_c$.

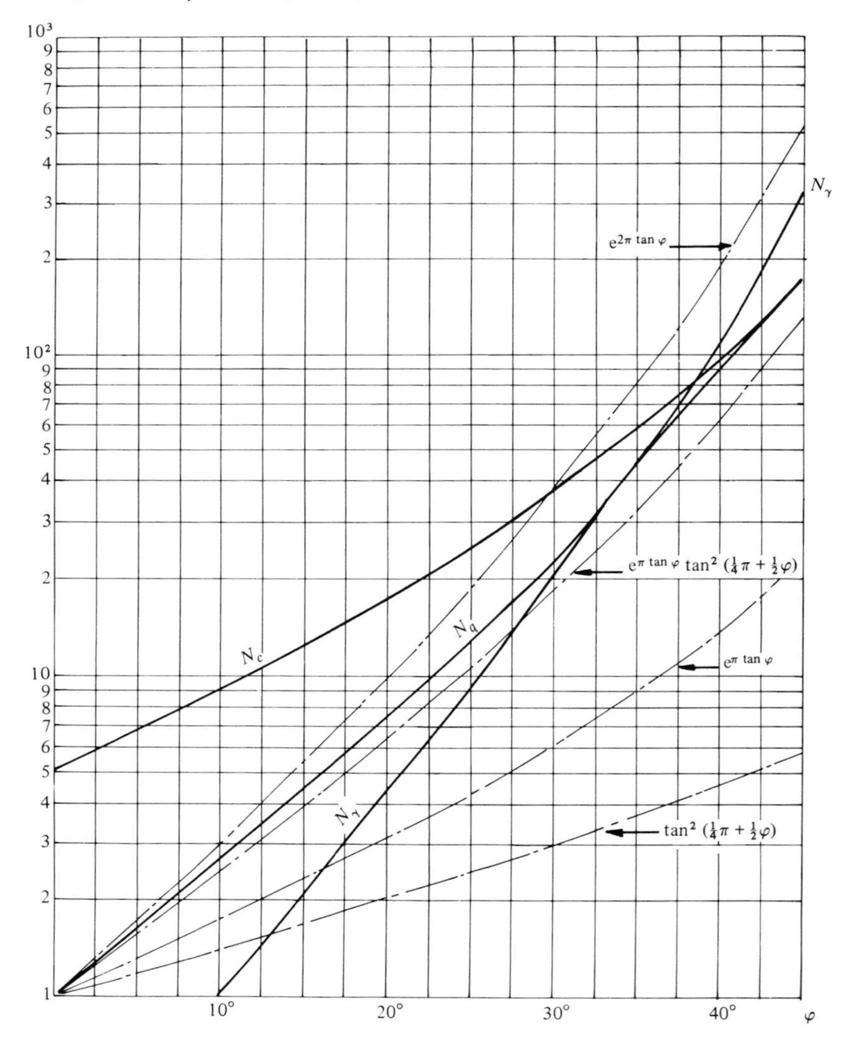

The particular values of N_c, N_γ, N_q for $\varphi = 0$ can be obtained by passing to the limit of the solution.

(2) When the rigid–plastic model is far from representing reality, especially when working, for example, with loose sands or soft clays capable of large deformations before reaching the yield limit, Terzaghi suggests lowering the value of the mechanical characteristics c and φ by $\frac{2}{3}c$ and $\frac{2}{3}\varphi$. The factors N_c, N_γ, N_q remain the same as in table 7.1, while naturally reducing the laboratory value of φ by a factor of $\frac{2}{3}$.

(3) Certain authors envisage different cases for different foundation footings: rigid foundation, flexible foundation, smooth-based foundation, rough-based foundation. In fact, for questions relating to bearing capacity these subtleties are only of academic interest: the base of an ordinary foundation should always be considered to be rough and rigid.

(4) In all the preceding equations, γ is the apparent density that is the bulk density, or the submerged density if the soil is below the water table or if there is any possibility of its becoming so.

(5) The safety factor should not be calculated from the weight of earth. The allowable bearing pressure is thus given by the equation:

$$q_a = [cN_c + \gamma D(N_q - 1) + \tfrac{1}{2}\gamma BN_\gamma](1/F_s) + \gamma D.$$

7.3.3 *Isolated foundation in the general case (three-dimensional problem)*

The general equation for a square foundation whose side is of length B is

$$q_u = 0.7\gamma(\tfrac{1}{2}B)N_\gamma + \gamma DN_q + 1.25cN_c.$$

7.3.4 *Foundation at the surface of a natural soil of limited thickness ($c \neq 0; \varphi \neq 0$)*

It is assumed that the layer of soil of thickness h rests on a more resistant medium that is not deformable, such as rock. The ultimate pressure under an undeformable foundation is given by

$$q_u = \frac{c}{\tan\varphi}\,\frac{h\cot\varphi}{B}\,e^{B/2a},$$

with

$$1/a = (2/h)\tan^2\varphi(\cot\varphi + \varphi + \tfrac{1}{2}\pi),$$

an equation valid for $B/2h > 3$ (Mandel, 1942). It is easily seen that q_u quickly reaches extremely high values when B/h increases.

7.3.5 *Inclined load*

If the resultant force makes an angle α with the vertical, an equation that accords well with reality can be found by kinematic considerations and by considering a net of characteristic lines close to the classic model (fig. 7.7)

$$q_i = (1 - \alpha/90°)^2 (cN_c + \gamma DN_q) + \tfrac{1}{2}(1 - \alpha/\varphi)^2 \gamma BN_\gamma.$$

The equations for bearing capacity for inclined load allow the load to be reduced to take into account seismic effects.

If for any reason, failure cannot take place in the direction of the shortest slip line (that is towards the left in fig. 7.7), for example because of an obstacle, or the presence of another construction, the inclination of the force increases the resistance since the failure can only occur in the direction of the longest slip line. The preceding equation gives the correcting factors if the sign of α is changed.

7.3.6 *Vertical eccentric load*

A foundation placed on the surface of soil cannot withstand tensile forces. In the case examined below it is assumed that eccentricity does not

Fig. 7.7. Inclined load:
(*a*) representation of failure in a two-dimensional model and
(*b*) model of plastic flow.

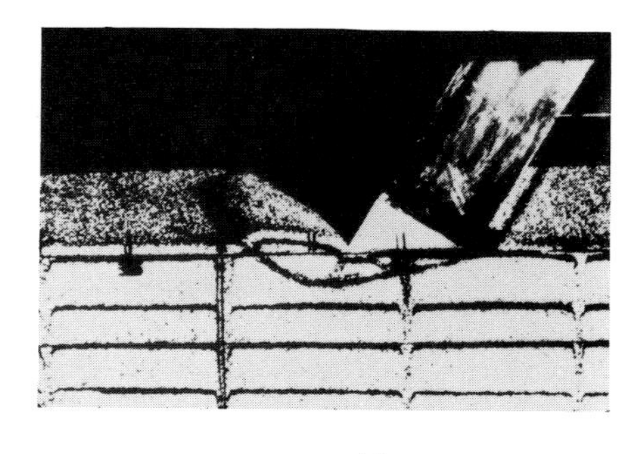

(*a*)

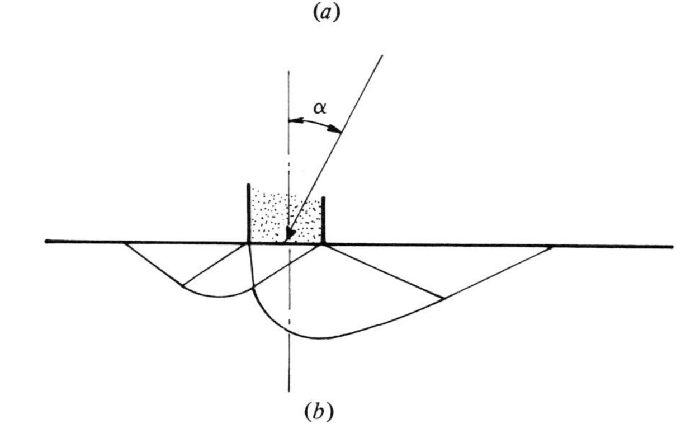

(*b*)

extend beyond the central third of the foundation. In the opposite case it is the eccentricity that defines the active breadth of the foundation which can result in a significant decrease in the actual value of B.

If the resultant of the forces has eccentricity $e < B/6$ (fig. 7.8), considerations analogous to those of the preceding paragraph lead to the equation

$$q_e = (1 - 2e/B)(cN_c + \gamma DN_q) + \tfrac{1}{2}(1 - 2e/B)^2 \gamma BN_\gamma.$$

If, for any reason, the slipping does not take place in the direction of the shortest line (that is to say towards the right in fig. 7.8) failure will occur towards the longest slip line with an increase in resistance. The preceding equation gives the correction factors by changing the sign of $2e/B$.

7.3.7 *Eccentric, inclined load*

The preceding correcting factors apply once again, but a rapid kinematic study is necessary to see if the inclination and eccentricity are combined or opposed.

7.3.8 *Influence of nearby foundations, asymmetries, etc.*

In practice, the presence of an isolated footing unconnected to the foundation studied does not modify the limit load. If a footing is linked to the foundation the influence is generally negligeable.

If the soil is asymmetric (presence of cellars, basements for example) the plastic model that gives the smallest load should be used. For example, D will be the depth of the foundation above the level of the deepest basement.

7.3.9 *Heterogeneous medium*

When the soil is heterogeneous and composed of layers whose resistance does not increase as a function of depth, care must be taken to ensure that the stresses caused by the presence of the foundation are acceptable in relation to the shearing resistance of the soil at every level (fig. 7.9). One assumes that stress

Fig. 7.8. Eccentric load: model of plastic flow.

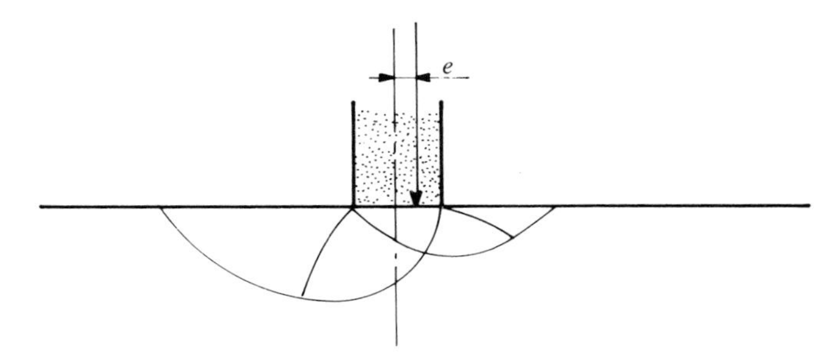

distribution is uniform in a 30° dihedral if the elastic modulus increases with depth or 45° if it remains constant or decreases, and one checks that at every level the bearing capacity equations are satisfied.

7.4 Safety factor

The safety factor used in building foundations is around 3 or 4. For concrete structures, the safety factor is, as we have mentioned before, more a factor of uncertainty. This is not the case for soil foundations, or at least the proportion of uncertainty is greatly reduced. The real role of the safety factor with regard to failure in soil mechanics is to establish a level of stress that only produces small deformations. If a structure is capable of withstanding large deformations, this factor can be lowered: as is the case, for example, with earth dams where the safety factor can be as low as 1.3 and even less in certain situations. It can, for example be 1.0 for the completion of a construction if the 'calculated risk' allows it, or 0.8 in certain cases for the hypothesis of the earthquake considered as a brief force. The idea of safety goes further than this. If very precise information on the mechanical characteristics of the soil is not available, or if the construction is of exceptional value, etc., increasing the safety might be a necessity. If it costs little to increase the depth or breadth of a foundation, then an insurance can be bought at very low cost. On the other hand, one can be less demanding about temporary construction work and accept the eventuality of disorders. Whatever the case, a distinction must be made between the safety factor (establish a level of stress lower than general failure,

Fig. 7.9. Punching on a layer of cohesionless material ($c = 0$) placed above a purely cohesive material ($\varphi = 0$) and resultant deformation (indentation).

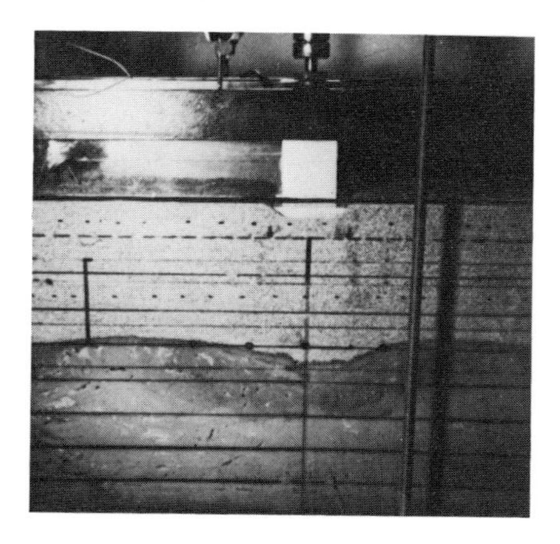

that will only cause small deformations) and the idea of safety (uncertainty about the mechanical characteristics and the size of acceptable risks).

7.5 Foundations in practice

The allowable bearing pressure for any situation that may crop up in current practice can be calculated by means of the equations in sections 7.1, 7.2 and 7.3. Obviously the parameters γ, φ and c have to be used, and these should always be correctly measured. Thus there is no point in carrying out consolidated undrained laboratory tests on a clay unless the time taken during construction and the geometry of the layers allow consolidation to take place. Here it is a question of good coordination between the designer and the laboratory (figs. 7.10 and 7.11). However it is interesting to look at the statistics of the most frequent accidents with shallow foundations that have taken place in France over the last 20 years (Logeais, 1971):

25% of the accidents are caused by foundations on fills the compacity of which is small (new fills);

20% are caused by unexpected water infiltration;

20% are caused by heterogeneous foundations, that is foundations whose deformations under load are unequal (isolated footings and strip footings, foundations at different depths, or isolated footings built at the same depth but which have different bearing capacities);

10% of accidents are caused by foundations not built to a great enough depth (frost, washing away);

10% of accidents are caused by settlement brought about by the construction of nearby buildings;

10% of accidents are caused by the foundations being built on very compressible soil (peat, soft clay, etc.);

5% of accidents are caused by building on unstable soils (slopes, quarries, mines, etc.).

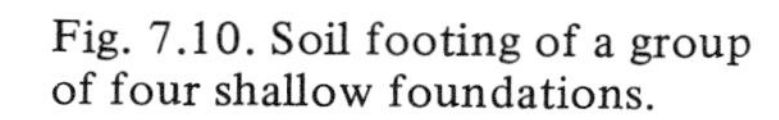

Fig. 7.10. Soil footing of a group of four shallow foundations.

Fig. 7.11. Isolated shallow foundations (north part of the experimental track of the aerotrain).

8

SETTLEMENT OF SHALLOW FOUNDATIONS

We saw in the last chapter how to calculate for most cases the limit load that leads to rupture of the soil beneath the foundation and how to determine the allowable bearing pressure using this value. But this approach is not always sufficient and one must still verify whether the possible delayed deformations are compatible with the proposed construction. The calculation of the settlement and its evolution in time, which are considered here, sometimes defines a limit which in effect makes it necessary to lower the allowable bearing pressure or at least to take a certain number of precautions in the construction itself.

When a foundation loads a soil without bringing about rupture an instantaneous deformation of the elastic type obviously takes place even if it is not reversible. This can be calculated by the theory of elasticity using the coefficients E and ν determined in the laboratory or in the field. The appearance of this deformation does not generally have any interesting consequences for the engineer since it appears simultaneously with loading at a time when the structure is not monolithic and can still be adapted in many ways. This is not the same as delayed deformations which can arise in the years following the completion of the construction. The quantity that is particularly important for engineers is the difference between the final deformation (at the end of a theoretically infinite period) and the initial deformation. This settlement comes about essentially as a result of the progressive expulsion of pore water from clayey soils by the mechanism of consolidation which we have already encountered when defining the viscoelastic properties of soils.

Identical phenomena occur at the moment of unloading and the methods of calculation are the same even if the elastic coefficients are different. In particular with earthworks of the order of 500 000 m^3 or more, such as an excavation for car parks or foundations of very tall buildings, for locks or certain hydroelectric works, the instantaneous swellings pass virtually unnoticed; however when they are measured (Mathian, 1971) one discovers that movements of some 10 to 20 cm take place, which are subsequently followed by delayed swellings which stop, then reverse, as a function of the progress of construction and subsequent loadings.

8.1 Calculating the settlement of a clayey layer confined between two permeable layers

It is assumed that the compressible layer is thin enough to be subjected to a uniform field of stress. Its thickness is $2H$. It lies between two permeable layers that ensure perfect drainage. This model, known as one-dimensional consolidation, corresponds exactly to that of the oedometer; it closely approximates a good number of practical cases, and the application of the calculation will be all the more satisfactory the closer the reality is to these hypotheses. The clay is characterised by a consolidation diagram (fig. 8.1) such that when the pressure increases from p_1 to p_2, the void ratio decreases from e_1 to e_2, and over a sufficiently small interval the diagram can be taken as a straight line (in natural coordinates). After the load p_2 has been applied for a time t, the void ratio becomes e; with the result that the effective stress is equal to p and consequently a pore pressure develops, $u = p_2 - p$.

U_z is called the degree of consolidation and is defined:

$$U_z = \frac{e_1 - e}{e_1 - e_2}.$$

The slope of the compressibility curve $a_v = -(\mathrm{d}e/\mathrm{d}p)$ is a coefficient characteristic of the soil. One has

$$U_z = \frac{e_1 - e}{e_1 - e_2} = \frac{p - p_1}{p_2 - p_1} = 1 - \left(\frac{u}{p_2 - p_1}\right).$$

As a result of the pore pressure gradient water tends to escape in a movement governed by Darcy's law (section 1.5), so one can write

$$V = (k/\gamma_w)\ \mathrm{grad}\ u,$$

where V is the velocity of water exterior to the soil sample and k is the coefficient of permeability. The quantity of water that flows in a volume element of soil $S\,\mathrm{d}z$ corresponds to the variation in volume of this element. It is the difference between the water which leaves by the face at z and that which

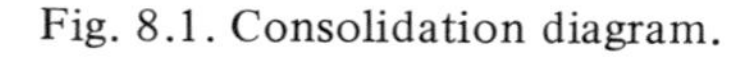

Fig. 8.1. Consolidation diagram.

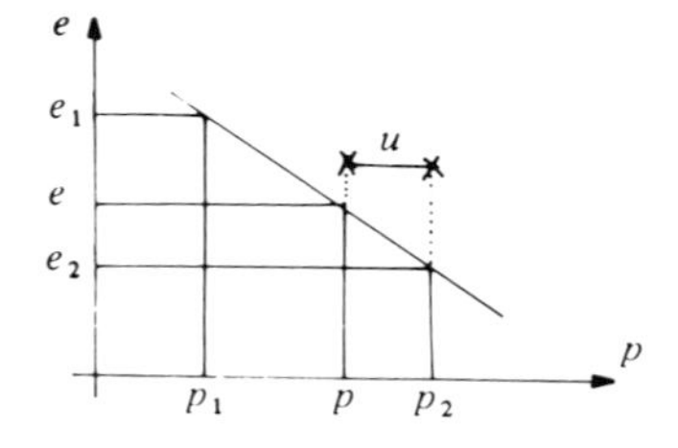

enters by the face at $z + \mathrm{d}z$.

$$\frac{\partial}{\partial t}(S\,\mathrm{d}z) = S\,\frac{\partial V}{\partial z}\,\mathrm{d}z = S\,\frac{k}{\gamma_w}\,\frac{\partial^2 u}{\partial z^2}$$

At the same moment the rate of variation of volume can be defined as the rate of variation of volume of the voids $S\,\mathrm{d}ze/(1+e)$, for the minerals are practically incompressible. Since $S\,\mathrm{d}z/(1+e)$ is the volume of solid minerals thus constant, the variation in volume of the voids can be written $(S\,\mathrm{d}z/(1+e_0))\mathrm{d}e$. Since

$$\mathrm{d}e = -a_v \mathrm{d}p = a_v \mathrm{d}u$$

then

$$\frac{k}{\gamma_w}\,\frac{\partial^2 u}{\partial z^2} = \frac{a_v}{1+e}\,\frac{\partial u}{\partial t},$$

and writing

$$c_V = \frac{k(1+e)}{a_v \gamma_w},$$

one has

$$c_V\,\frac{\partial^2 u}{\partial z^2} = \frac{\partial u}{\partial t},$$

which has the classic form of the heat equation. The boundary conditions of the consolidation problem are:

$$u(t) = 0, \qquad \text{for } z = 0,$$
$$u(t) = 0, \qquad \text{for } z = 2H,$$
$$u(t) = p_2 - p_1, \qquad \text{for } t = 0.$$

The classic solution to this problem is:

$$U = 1 - \sum_{m=0}^{m=\infty} \frac{2}{M^2}\,\mathrm{e}^{-M^2 T}$$

with

$$M = \tfrac{1}{2}\pi(2m+1)$$

and

$$T = c_V t/H^2$$

called the time factor. The numerical values of this solution are shown below.

U	0.1	0.2	0.3	0.4	0.5	0.6	0.7	0.8	0.9
T	0.01	0.03	0.07	0.13	0.20	0.29	0.40	0.57	0.85

It can be seen from these results that the duration of consolidation is proportional to the square of the thickness of the layer.

The time of the half-settlement, that is $U = 0.5$ where $T = 0.2$ gives an order of magnitude for the length of time in which principal disorders may occur. One has:

$$t(\text{half-settlement}) = \frac{\gamma_w a_v H^2}{5k(1+e)}.$$

From the point of view of evolution in time nothing changes if at the initial moment the pore pressure u varies linearly with depth rather than being rigorously constant, the mean pressure simply replaces the constant $u = p_2 - p_1$.

Let us finish by noting that if the compressible layer rests on an impermeable surface, one only has to substitute its thickness H into the preceding case for it to remain valid (fig. 8.2).

8.2 Secondary settlement

Another delayed phenomenon, known as secondary settlement, exists alongside that of hydrodynamic or primary settlement. This corresponds to a flow of the solid mineral skeleton of the soil. It follows an approximately linear law as a function of the logarithm of time. That is to say, it continues to manifest itself for a long time after the hydrodynamic settlement has ceased.

This phenomenon is particularly evident in peats and recently formed muds.

Fig. 8.2. Degree of consolidation in a layer of thickness $2H$ as a function of the depth and duration of loading.

U_z = degree of consolidation (%)

$$T = \frac{c_V t}{H^2} = \frac{k(1+e)}{a_v \gamma_w H^2} t$$

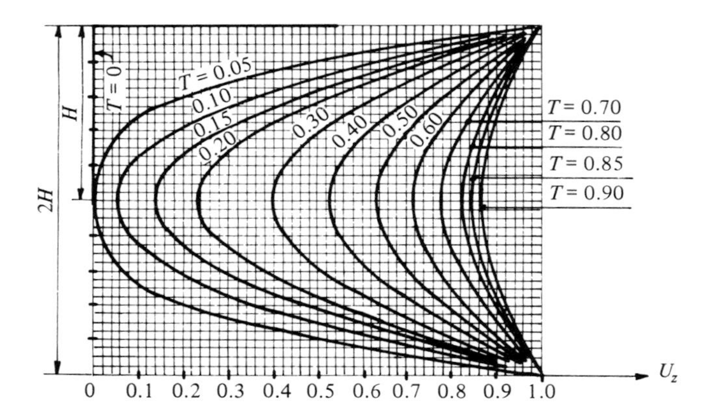

It is less evident in the modern or glacial clays found in France and is practically non-existent in French tertiary or even older clays. Secondary settlement takes place more or less without any pore pressure (since the process is very slow) and it can be found by simple analogy from the results of compressibility tests. This phenomenon is of little importance in the majority of the soils of metropolitan France.

8.3 Distribution of stresses in the soil

The theory of consolidation outlined above uses the hypothesis of a uniform field of stress. In reality it is not uniform beneath a foundation. The stresses decrease gradually the further away they are from the footing. The distribution of stresses has been studied by numerous authors using the theory established by Boussinesq for a single point load applied at the surface of a semi-infinite medium. Boussinesq gave the expressions for the tangential, radial, vertical stresses ($\sigma_t, \sigma_r, \sigma_z$) and for the shearing stress τ_{rz} (see fig. 8.3) at every point, with the most remarkable result that the expression σ_z is independent of Poisson's ratio (while the others are not). Using the notation of fig. 8.3,

$$\sigma_z = \frac{3P}{2\pi}\frac{z^3}{\rho^5} = \frac{3P}{2\pi z^2}\cos^5\theta.$$

The meridians of the surfaces σ_z = constant have a circular appearance and are tangential to the surface of the soil at the point of application of the force P. From the stresses produced by an isolated force one can compute, by fairly complicated integration, the influence at a point of a load distributed arbitrarily over a given surface. σ_z is still independent of ν of course.

At the moment there are some explicit solutions available for a certain number of load problems; for a long time there have been convenient graphical solutions for other cases. As early as 1935 Newmark developed influence charts which are easy to use. Below is an example of an explicit solution for the distribution of stresses within a flexible circular surface of radius R uniformly

Fig. 8.3. Point load at the surface.

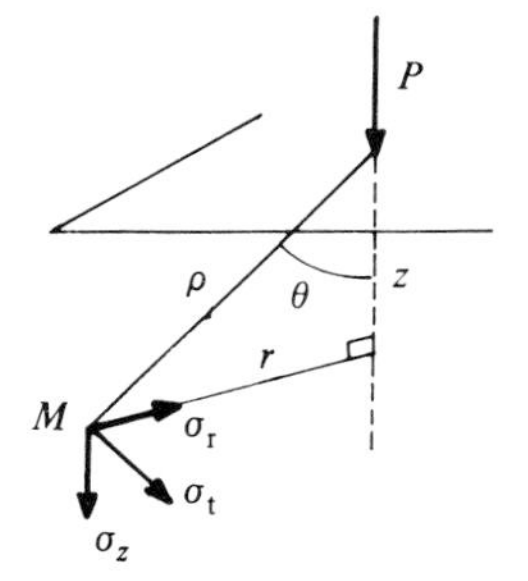

charged at a pressure p:

$$\sigma_z = p\left[1 - \left(\frac{1}{1 + (R/z)^2}\right)^{3/2}\right].$$

Boussinesq's theory also gives the soil deformations under load. The vertical displacement produced at point M by the load P is

$$w = \frac{P(1+\nu)}{2\pi E\rho}\left(2 - 2\nu + \frac{z^2}{\rho^2}\right),$$

in particular the displacement at the surface ($z = 0$), where r is the radial distance from the point of application of the force, is

$$w = \frac{P(1-\nu^2)}{\pi E r}.$$

This expression is obviously not valid at the point of application of the force.

The explicit calculation of the displacements under the flexible circular footing results in the following values:

$$\text{at the centre} \quad w_0 = \frac{2P(1-\nu^2)}{\pi E R} = \frac{2pR}{E}(1-\nu^2);$$

$$\text{at the edge} \quad w_e = 2w_0/\pi.$$

Under a rigid circle Boussinesq has shown that the distribution of pressure is

$$p(r) = p/2[1 - (r^2/R^2)]^{1/2},$$

Table 8.1. *Values for C_f.*

	Centre	Edge	Average
Flexible circle (R = radius)	2	1.27	1.70
Rigid circle	1.57	1.57	
	Centre	**Corner**	**Average**
Flexible square (R = half side)	2.24	1.12	1.90
Rigid square	1.76	1.76	
		Corner	**Average**
Flexible rectangle (R = half breadth)			
Length/breadth = 2		1.53	2.60
3		1.78	3.05
5		2.10	3.66
10		2.58	4.50

(with $p\pi R^2 = P$); the displacement of the whole is equal to the displacement at the centre, thus one has

$$w = 1.57\,pR(1-\nu^2)/E,$$

which is a value close to the average settlement of the flexible, uniformly loaded circle. In a general way one can write

$$w = C_f pR(1-\nu^2)/E,$$

where C_f is a coefficient that is determined by the point chosen, by the shape, and by the stiffness of the footing. Table 8.1 gives values for C_f from which the majority of practical cases can be resolved.

Using these values one can find by sum of the deformations at the vertices of two or four rectangles, the deformation at one point on the perimeter or the deformation at a point on the surface of the footing. One can also easily find by sum and difference the deformation of a point at the exterior of a flexible loaded rectangle.

8.4 Calculating the settlement beneath foundations

There are two possible methods using Boussinesq's theory. The first is a global method and consists of estimating the deformations from the elastic modulus and Poisson's ratio. It is used mainly for homogeneous soils when the elastic coefficients are truly constant. The second method is to determine the distribution of stresses at each level beneath the foundation and to divide the soil into a certain number of layers (more or less imaginary) making use of the local geology, in such a way that one can apportion to each a mean value σ_z having a physical significance, that is to say such that σ_z varies only slightly in a layer. The settlement of each layer is then calculated. If e_0 is the initial void ratio of the soil subjected to the weight of earth above it, and e_1 the void ratio after application of the surcharge (values being given by the compressibility test), the settlement of a layer of thickness Δh will be:

$$w = \Delta h \frac{e_0 - e_1}{1 + e_0}.$$

The settlement of the different layers is then summed. The evolution of the settlement in time is determined by the thickest or the most impervious layer. This method of cutting into slices obviously yields only fairly approximative results, to the extent that it is mostly used in stratified media, when the theory for the distribution of stresses becomes less accurate. Further, it is assumed that none of the layers expand laterally, which is an incorrect assumption since one is far removed from the theories of one-dimensional consolidation. These latter considerations highlight the advantages of the global methods which have been developed for a few relatively simple heterogeneous media with one or two layers, whether the solutions are explicit or given by charts (Giroud, 1972).

8.5 Allowable settlement of foundations

In general the uniform settlement of a structure does not present any significant danger and does not bring about major accidents. Differential settlement between footings can, on the contrary have totally catastrophic effects (figs. 8.4 and 8.5). The extent of the allowable settlement depends on the nature of the constructions (metal framework, masonry, brick infill, reinforced concrete, prestressed concrete, isostatic or hyperstatic systems, etc.). One can

Fig. 8.4. Differential settlement: the lamppost at the right indicates the vertical.

Fig. 8.5. Differential settlement: the Guadalupe basilica in Mexico.

nevertheless, from experience, adopt a certain number of rules established from observations in the field. For example, it has been established that the differential settlement of isolated foundations on sand or clay can reach 60% of the maximum settlement under a footing; for raft foundations it is generally smaller, of the order 35–60%, both on sands and clays. On the other hand, experience shows that a structure can support differential settlements of the order of 1/300 of the distance between footings without apparent effect.

In standard buildings another limit can be given for differential settlement: if it is greater than 5 cm, fissures and ruptures always appear in structures, thus making the placing of joints indispensable. It should be pointed out that fissures appear more easily when the soil is a sand rather than a clay. Lastly, if one takes into account the uncertainty of the calculation of settlements (which can be estimated to be at 50%), one arrives at the following rules which should be observed when setting up a project:

Allowable differential settlement (angle of rotation)	1/500 of the reach between footings (isolated footings or raft foundations) 1/1000 when one requires maximum certainty	
Allowable differential settlement (size)	Clay: 3–4 cm Sand: 2–3 cm	
Total settlement (size)	Isolated foundation	Clay: 6 cm Sand: 4 cm
	Raft foundation	Clay: 10 cm Sand: 6 cm

It goes without saying that the preceding values have been obtained for normal structures in standard conditions. They do not take into account the very nature of the constructions, whose rigidity and strength can vary from one type to another: they should simply be considered as giving an order of magnitude for standard buildings.

8.6 Elastic instability of foundations

When a very tall building rests on an elastic soil, instability may occur that is allied to buckling.

Let P be the weight of a structure whose centre of gravity, G, is at a height h above the foundation plane. Under the effect of an external force F (for example due to the wind), which is assumed to be applied at a height h, the construction leans at an angle θ which brings about an elastic reaction at the soil level and a restoring couple C (fig. 8.6). The equilibrium equation allowing calculation of θ is written

$$Fh + Ph\theta = C\theta, \tag{8.1}$$

from whence

$$\theta = Fh/(C - Ph). \tag{8.2}$$

If F tends towards zero, (8.1) defines a critical height for the centre of gravity, $h_c = C/P$, such that an external pertubation, however small, causes the structure

Fig. 8.6. Elastic instability, point G is the centre of gravity.

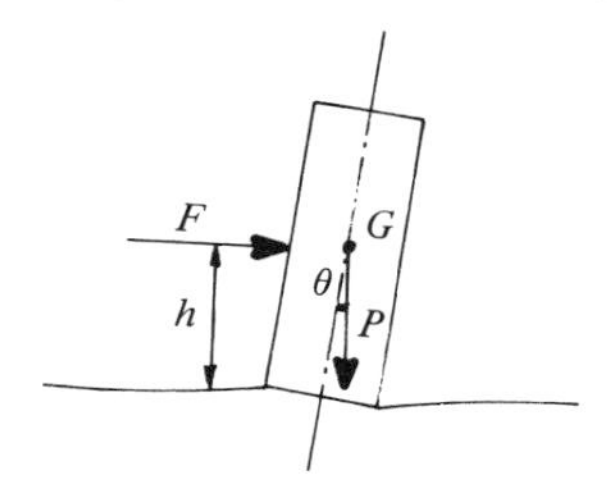

Fig. 8.7. Long-term instability: the leaning Tower of Pisa.

to tip. When this value is incorporated in equation (8.2) it can be seen that θ becomes infinite which signifies that equilibrium is no longer possible.

The theory of elasticity allows C to be computed explicitly when the rigid foundation has a circular base of radius a:

$$C = \tfrac{4}{3} Ea^3/(1-\nu^2).$$

As we have seen elsewhere the settlement w under the weight P is

$$w = \frac{P}{2a}\,\frac{1-\nu^2}{E},$$

thus

$$h_c = 2a^2/3w,$$

which eliminates the parameters that are not always well known, P, E, or ν, leaving the quantities a and w which are, on the other hand, directly measurable. The linear elasticity of the soil is, however, assumed in this formula. It does not differentiate between instantaneous deformation (short-term instability) and delayed deformation (long-term instability, fig. 8.7). With the characteristics of instantaneous deformation one can calculate the period of oscillation of the construction to arrive at the following general formula:

$$\omega = \frac{2\pi}{T} = \frac{(3g)^{1/2}}{2h}\,(h_c - h)^{1/2},$$

which shows notably that the critical height h_c is that which corresponds to an infinite period.

Finally, in certain conditions one should consider increasing the safety factor of foundations of tall constructions (Habib and Puyo, 1970; Habib and Luong, 1972).

9

BEARING CAPACITY AND SETTLEMENT OF DEEP FOUNDATIONS

When the load borne by a structure is very large, or when the mechanical qualities of the soil are poor, it is necessary to consider the use of deep footings. Deep foundations, piles, shafts, caissons, continuous walls, etc. can be defined simply (as opposed to surface or shallow foundations) as foundations where the term $\frac{1}{2}B\gamma N_\gamma$ of the general formula (see chapter 7) is negligeable.

From the point of view of the theory of bearing capacity, practically no distinction can be made between piles, shafts and caissons. But it is clear that there is, in practice, a considerable difference between a shaft excavated by hand, and a mechanically driven pile, a difference that is not wholly due to the size of the cross-section. Besides, nowadays with the possibility of constructing piles larger than ordinary shafts, this difference is disappearing. As for caissons, their transverse dimensions, and consequently their heights, are so large that the local stratigraphic conditions make the use of equations or rules established for a semi-infinite continuous medium very misleading. And finally, the continuous wall, a true wall cast at depth, is a problem in two dimensions. In France the usual function of deep foundations is to seek out a good soil, but the complementary bearing capacity of the layers crossed is not negligeable. Floating piles is the name given to those piles which do not cross any layers of sufficient strength to be supported at the tip, but are instead supported by skin friction.

9.1 Pile technology

Since the dawn of lake-dwelling civilisations the industrial genius of man has delighted in inventing new pile devices. There are a considerable number of piles that cannot possibly all be cited here; we can, nevertheless, attempt to classify them.

9.1.1 *Driven piles*

These are installed by the blow of a hammer (mechanical, diesel or vibratory, etc.) to the rectilinear element of wood (not often today), concrete or iron. The mass of the hammer should be adapted to the mass of the pile; if it is too light it is ineffective, if it is too heavy it destroys the head of the pile despite the use of pile caps. Concrete piles, the type most frequently used in France, are prefabricated (so that the concrete becomes sufficiently hard), and their length is not necessarily right. Thus once it has been installed if it is too long it is cut, if too short it is spliced. Metallic piles are employed relatively infrequently in France, at least in the form of a T or H section. On the other hand driven metallic pipes, with a loose shoe, are used as formwork for cast-in-place piles, the casing being recovered afterwards. Generally speaking, driven piles do not easily accommodate heterogeneities in the soil (lumps, nodules, pieces of masonry, old foundations, etc.) which cause deflections of the pile and ruptures in the shaft.

9.1.2 *Cast-in-place piles*

With this technique the walls of a hole serve as a formwork for concrete which is poured into it. A temporary prop is used either as a means of opening up the hole (loose shoe), or to make the opening of the hole easier (casing) (see figs. 9.1 and 9.2). Recovering the pipe is a delicate operation: it must be done when the concrete is still wet avoiding at any cost constriction of the pile or dilution of the cement grout by a rising hydraulic gradient. The operation is carried out by ramming the bottom of the hole (Franki), by a tacking action in the shaft (Benoto), or by energetic vibration, etc. The shape of the shaft of these piles is not very regular, which ensures a firm lateral grip, even though concrete swelling occurs where the soil layers are softest.

In bored piles where the excavation techniques are derived from those used in oil wells, heavy drilling mud ensures that the hole is supported. At the moment

Fig. 9.1. Pile excavation using a rotating bucket machine (Caldwel – 100 hp).

Fig. 9.2. Concreting of a cast-in-place pile.

of concreting, the seeping concrete displaces the mud by gravity and provides effective cover of the metallic reinforcements. The shape of the piles is, irrespective of the soil qualities, very regular. With these techniques it is possible to enlarge the foot of the pile (in a bulb form) to increase the bearing capacity at the point. Very great depths (up to 60 m) and large diameters can be attained by this method of working. The stress in the concrete of a pile is never above 5 to 7 MPa.

9.1.3 *Other piles*

These include piles that are installed by water-jets applied at the foot (jetting), jacked piles, generally with short elements for underpinning (Méga), screw piles with large thread and small helices (for installation), screw piles with small thread and large helices (for marine construction), raking piles, piles to withstand tractive forces, piles grouted at the foot so that the base is consolidated, concrete piles mixed with soil if it is coarse enough, sand piles, etc.

9.1.4 *Shafts*

Shafts are manually excavated: their diameters vary from about 0.80 to 2 m. Their chief advantage lies in the fact that they permit direct observation of the terrain. In certain regions of France, and on small sites, skilful labour can, by this traditional method, compete with the modern methods described above which call for heavy plant to be brought in.

9.1.5 *Caissons*

Whether made from metal or concrete the most frequent method of installing caissons is undercutting, more or less helped by loading, embanking, jetting, etc.

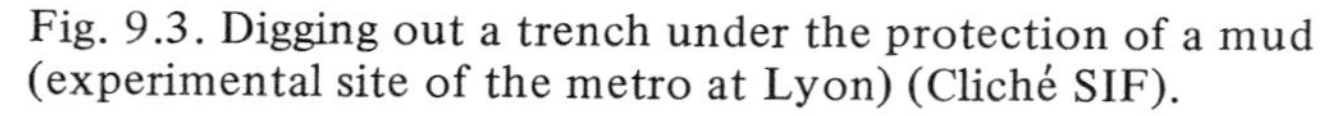

Fig. 9.3. Digging out a trench under the protection of a mud (experimental site of the metro at Lyon) (Cliché SIF).

9.1.6 *Cast in situ diaphragm wall*

An excellent modern method, only recently developed, the cast *in situ* diaphragm wall is firstly a retaining structure rather than a foundation, and in this role it is redundant (figs. 9.3 to 9.7). If the mode of execution varies with the means of the contracting firm, the principle always remains the same. A

Fig. 9.4. Installation of the reinforcement for a cast *in situ* diaphragm element (experimental site of the metro at Lyon) (Cliché SIF).

Fig. 9.5. Two concentric cast *in situ* diaphragm walls forming the housing for the skirting of a gasometer.

ditch is excavated under the protection of bentonite stabilised mud, which can be made heavy or over-pressurised in difficult cases, and the wall itself is made from elements between 3 and 15 m long (in exceptional circumstances they can be 20 m long) and the different panels are connected together to ensure that the wall is watertight. If one requires accuracy of installation at great depth (as deep as 50 m or more) then obviously perfect precision is demanded of the excavating

Fig. 9.6. Large embankment site protected by a peripheral cast *in situ* diaphragm wall.

Fig. 9.7. Cast *in situ* diaphragm wall supported by vertical legs and two lines of ground anchors (Boucicaut car park, Paris).

technique. The state of the surface of the diaphragm wall is rarely perfect: it depends on the nature of the soil and the deviations of the excavation, and sometimes necessitates a significant smoothing layer. To avoid this work, especially when an outline has to be followed, one can use prefabricated elements of reinforced concrete which clamp on to each other like veritable concrete sheet-piles; they are submerged in a trench filled with excavating grout and an admixture that ensures the setting of the cement, or with a mud for which a cement grout will be substituted. This technique requires fairly large hoisting apparatus.

9.2 Bearing capacity of deep foundations

It is common nowadays to separate the bearing capacity of a pile into a force at the point and skin friction in order to take into account the considerable disproportion that exists in most cases between these two factors (floating piles or piles which have reached good soil). But, aside from this consideration, it is clear that this simplification is excessive, and that the two types of resistance cannot be summed. Proof of this can be seen in the fact that one can, in the region of the ultimate bearing capacity, transfer a significant part of the lateral load to the load at the point and vice versa (Kérisel's experiment). In the same way a pull-out force only gives an approximate idea of the skin friction resistance, an idea which is closer to reality the closer the angle of internal friction is to zero. Elsewhere the conceptions from the theory of plasticity with Coulomb's material are only a small help, since, as we have seen for shallow foundations, these only resolve two-dimensional problems while all evidence for piles points to an axisymmetric problem; the idea of a

Fig. 9.8. Driving a pile starting from different depths.

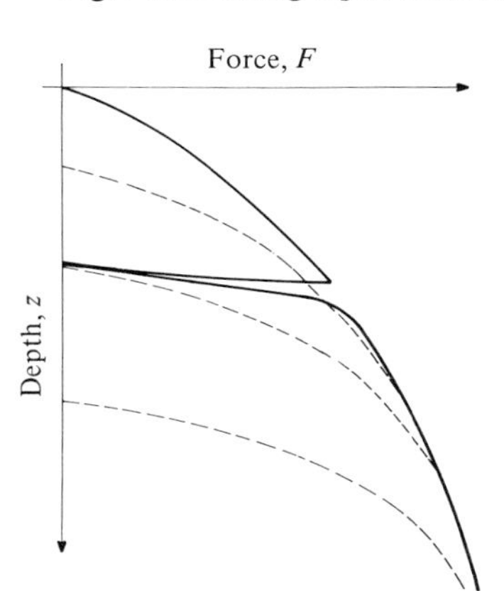

coefficient of shape is essentially empirical since it is based upon too few experiments to be used effectively.

Finally, the soil properties of elasticity and compressibility are probably large enough for the net of plastic slip lines in the soil not to emerge at the surface, but to correspond to a limited volume of restrained plasticity.

It seems that at the moment engineers are somewhat unconvinced by pile theory – this perhaps explains why certain regulations in countries other than France completely challenge the use of static formulae in a project. In fact there is a real difficulty in interpretation of pile tests, but it is surmountable if the pile indentation mechanism is understood.

When a pile is sunk in the soil from the free surface (fig. 9.8) the graph of the bearing capacity as a function of depth gives the resistance envelope of piles driven in the same medium, but starting from a certain depth. There is thus a difference between the bearing capacity of a pile driven in by a hammer or jack to a certain depth, and that of a pile cast in the soil to the same depth (and not driven). This statement cannot be considered as a value judgement between the two types of pile, for the only justification here is the lowering of foundation costs.

In marls and clays hammer hardening is quickly achieved. In sands and soils with high internal friction, on the contrary the hammer hardening necessary during installation is considerable; as penetration of several diameters from the initial position to regain the bearing capacity is needed. To express these two states it has been proposed that when estimating the end bearing capacity a depth factor be used:

$N_q = j e^{\pi \tan \varphi}$ for the initial resistance;

$N'_q = j e^{2\pi \tan \varphi}$ for the bearing capacity after hardening.

These two expressions correspond, in fact, to the clearly differentiated nets of slip lines. The values given above are hardly intellectually satisfying but they do provide a useful estimation for the preliminary design.

The skin friction also poses delicate problems; the slip surface is *a priori* simply defined. Unfortunately, except where $\varphi = 0$ the fact that one does not know the stress normal to the pile is a definite obstacle. It must also be remembered that when there is a clear peak in the stress/strain curve, there is no reason why the slipping should be the same beneath the point and along the shaft: it is not always correct to use the same values of φ and c for the skin friction and for point bearing capacity in a real homogeneous soil.

These difficulties of choice of mechanical parameters disappear if one can carry out static tests in the field on full size piles, or on small piles, or with penetration apparatus. In the latter case if the experimental apparatus allows one to separate the point bearing capacity and the skin friction, the total force on the penetrometer will be $F = Sp_1 + sp_2$ (S = lateral surface area of the shaft; s = surface area of the point; p_1 = mean lateral shearing; p_2 = mean pressure at the point), and the estimation of the bearing capacity of a pile at the same depth can be made using the following simple rule

$$F' = S'p_1 + s'p_2.$$

Experience shows that in clays or stiff marls and fissured materials this rule predicts increased skin friction and that it is prudent to lower the value of p_1 by a coefficient between 0.5 and 1. In the same way, for the end bearing capacity one must expect to adjust p_2 for the much larger absolute values of displacement for large piles than for penetrometers, that is to say that the ultimate pressure p_2 is greater for a pile with a small cross-section.

All these difficulties of interpretation show the advantages to be gained from a full-scale test (fig. 9.9); it should be added that a knowledge of the results of such a test allows one to use safety factors for failure of the order of 2 to 2.5,

Fig. 9.9. Installing the ballast for a pile loading test.

which are thus smaller than those used for shallow foundations on account of the work hardening, that is to say small deformations.

9.3 Bearing capacity of piles from dynamic tests

When the blow of a hammer on a pile produces a vertical penetration it is possible to deduce from it the resistance, R, to the displacement. In this sense driven piles are all 'tested' at the time of installation, at least comparatively. This is why it is important to keep a record card noting the displacement as a function of the number of blows; this displacement, or rather a tenth of the displacement corresponding to a series of 10 blows is called refusal.

Let M be the mass of the hammer driven at a speed v_0 and P the mass of the pile (fig. 9.10). Before the blow the momentum is: $Mv_0 + 0$. After the blow, if the hammer moves down with the pile (no elastic reaction) the combined velocity is

$$v' = v_0 M/(P + M).$$

If the strength of the soil is R this movement is uniformly slowed down with an acceleration: $\gamma = R/(P + M)$ from which the velocity is

$$v = v_0 \frac{M}{P+M} - \frac{R}{P+M}(t - t_0).$$

The pile stops ($v = 0$) for

$$(t - t_0) = v_0 M/R,$$

Fig. 9.10. Driving a pile.

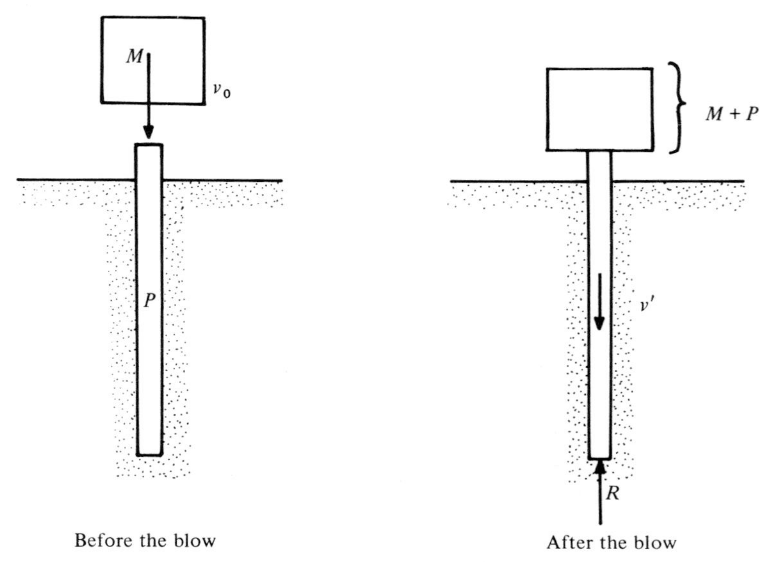

and the vertical displacement is then

$$e = \tfrac{1}{2}\frac{R}{P+M}(t-t_0)^2$$

Hence

$$e = \frac{Mv_0^2}{2}\frac{1}{R}\frac{M}{P+M}$$

or again, if h is the height of the fall of the hammer

$$R = \frac{1}{e}\frac{M^2gh}{P+M},$$

a formula known as the Dutch formula. It is generally used with a safety factor of 6.

The preceding formula shows the efficiency of the driving. Writing

$$Mgh = Re + Re\,P/M$$

(the second term representing the loss when the blow takes place) shows that the larger M, the greater the efficiency. In practice one is limited by the destruction of the pile by the blow but one should try to have $M > \frac{1}{2}P$. With diesel hammers one is limited to $M > \frac{1}{6}P$. Of course, when driving in a pile in order to measure the refusal, the cap is removed to limit the loss of energy due to the blow.

The Dutch formula is established in a very simple but not very rigorous way. It neglects, in particular, the experimental fact that there is a minimum height h_0, below which the blow does not produce permanent vertical displacement but solely elastic displacement. Hiley's formula, often used in England, is written:

$$R = \frac{h-h_0}{e}\frac{M^2g}{M+P} \quad \text{(safety factor: 3 to 4).}$$

Another formula that takes into account the elastic shortening e_1 of the pile is used in the United States; this is Crandall's formula:

$$R = \frac{h}{e+\frac{1}{2}e_1}\frac{M^2g}{M+P}.$$

Numerous other driving formulae exist taking into account more detailed and complex parameters, but since a choice has to be made, one can say that all the formulae are equally good or equally bad, for the major reason that the dynamic resistance does not have any great correspondence to the static resistance, except perhaps in sands and gravels. In the dynamic test the pore water does not have time to escape; in the static test consolidation allows other stresses to develop. This is demonstrated in the fact that often, when resuming driving piles after several days or weeks in the ground, they have an improved refusal. This fact is clearly illustrated in marls; it is perhaps amplified by thixotropic phenonema, but in any case the dynamic resistance and the maximum static load do not correspond. On the other hand, any driving formula whose safety factor is checked by a static test in a given site supplies an excellent means of comparison and control.

9.4 Negative friction

When a pile crosses a compressible layer before encountering a resistant soil, and if this layer settles, for example because of the surcharge of a recent surface fill, the effect of the skin friction can be reversed: instead of contributing to the bearing capacity of the pile it works to load the point. This phenomenon, which occurs in the compressible layer and those above it, is called negative friction.

Aside from the effect of negative friction, the placing of fills on inferior soil into which piles have already been driven is a very dangerous operation. The fills are rarely symmetrical, and the shafts of the piles are horizontally stressed, by forces which have not been accounted for. This type of rupture is all the more unfortunate because the piles are broken one after the other, and their transverse resistances cannot be added up.

9.5 Anchorages

When a ground anchor is made of iron bars or grouted cables in a vertical borehole the resistance of the anchorage is, as for the pile, an axisymmetric problem. As for the tie-bars which support continuous diaphragm walls, the direction of the axis of the ground anchor does not coincide with the vertical (fig. 9.11) and the problem becomes even more complicated. The force necessary to uproot a ground anchor arises from the shearing resistance mobilised along the lateral surface of the fixed anchor length in addition to the passive

Fig. 9.11. Excavation for an angled anchorage.

resistance exerted on the mass by the master couple due to irregularities in the anchoring bulb. These latter depend on the nature of the soil and the mode of execution; that means that they are generally not well known while their effect is far from being negligeable. The precise calculation is thus often illusory and it is advisable to test out the ground anchors before they are put to use; it is all the more simple if the anchors are prestressed (fig. 9.12). When one is planning to use a large number of ground anchors it is often economical to carry out one or several destructive tests to estimate, from the failure load, the load that can be supported. In such a case the safety factors used vary between 1.5 and 2 depending on whether the ground anchors are temporary or permanent (Bureau Sécuritas, 1977).

The real problem with permanent ground anchors is their durability: in the long term a release of stress can be caused by a creep of the soil or steel, or the tendons can corrode. The first danger can be eliminated by reducing the load

Fig. 9.12. Cross-section of a tie-bar with prestressed injected cables.

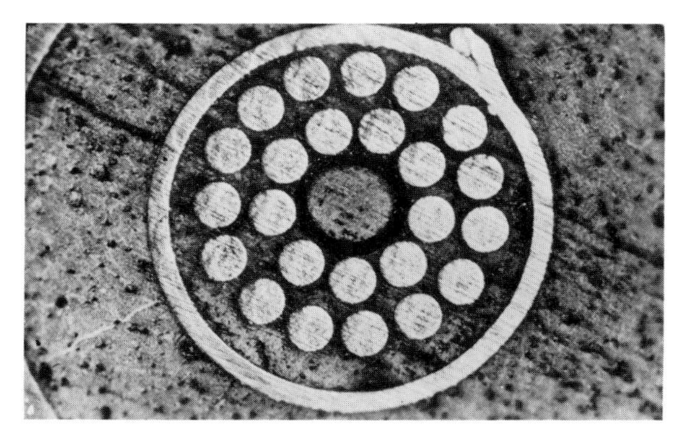

Fig. 9.13. Device for checking the tension of cables.

supported by the anchorage or the safe bearing capacity of the steel. One can try and eliminate the second danger by carefully protecting the metal from contact with the soil water. In all cases one can statically check the tension of the cables by the use of a permanent dynamometer which can detect an unexpected development due to the age of the construction or modifications to the nearby area (Habib and Puyo, 1972) (see fig. 9.13).

9.6 Grouped piles – settlement of piles

The bearing capacity of a group of piles is not necessarily the product of the bearing capacity of one pile and the number of piles. Introducing a large number of piles into loose sand can increase the compactness so that the quality of the soil is greatly improved in such conditions: the bearing capacity is thus improved as a result of the group effect. In other cases it is quite possible that the bearing capacity of the volume surrounding the piles is less than the sum of the individual bearing capacities: there is a general failure. This case can, in particular, arise when there is a soft layer beneath a thin resistant layer which had given a satisfactory refusal in driving. Each particular situation should thus be examined in its geotechnical context; in principle, one should try to place the piles at least two diameters away from each other (i.e. three diameters from axis to axis).

Studying the settlement reveals even more clearly the difference between the pile and the group, and it can be said that knowledge of the settlement of an isolated pile is of no real use to the engineer. In any case, it is less than the settlement of a group identically loaded. With a group of piles, the load of the construction is considered to act as a surcharge at the level of the pile points, and is decreased by the lateral friction of the external piles, a friction which is distributed to the deeper layers in a cone shape whose angle is only 10° to 20° at the vertex. The settlement can then be calculated using the same hypotheses as those used for surface foundations.

9.7 Accidents with deep foundations

As for surface foundations we shall conclude this chapter on the bearing capacity of foundations by looking at the simplified statistics of the accidents occuring to deep foundations in France (Logeais, 1981):

25% of accidents occur when a geotechnical survey has not been carried out;

35% of accidents result from poor interpretation of borings or from ignorance about the laws of soil mechanics;

15% are due to mistakes in execution;

10% come about as a result of failure of materials (rotting of wood at the water table, corrosion by water, etc.).

10

EARTH DAMS

10.1 Further calculations of slope stability

The methods used for calculating the stability of slopes which we discussed earlier apply in their entirety to the study of the stability of earth dams. However, although water is always suspected of playing a role in landslides, cases of failure in slopes may be encountered where this role is minor. In earth dams, on the contrary, the permanent presence of water in the reservoir, as well as the asymmetry of the upstream face (where the water filters in) and the downstream face (where the infiltration seeps out), necessitate the careful examination of the effect of the pore water.

10.1.1 *Revision of laminar hydraulics*

Let us study the forces acting on the solid elements of a soil subject to water percolation (fig. 10.1). The small element, the volume of which is $\mathrm{d}V$, between the two equipotentials AD and BC and the two flow lines AB and DC is in equilibrium under the effect of the reactions on the four faces. The forces on the element $ABCD$ itself are partly due to the weight of the submerged grains, that is to say $\gamma_i \mathrm{d}V$, and partly to the forces due to the loss of head correlating to the flow, that is to say between the elements AD and BC which is $\mathrm{d}h\ \mathrm{d}S$ ($\mathrm{d}h$ is the loss of head between A and B; $\mathrm{d}S$ is the cross-section of the flow tube). As $\mathrm{i} = \mathrm{d}h/\mathrm{d}s\gamma_\mathrm{w}$ (see section 1.5) where $\mathrm{d}s$ is the distance traversed, one can write

$$\gamma_\mathrm{w}\ \mathrm{i}\ \mathrm{d}s\ \mathrm{d}S = \gamma_\mathrm{w}\ \mathrm{i}\ \mathrm{d}V.$$

Fig. 10.1. Weight and force associated with the flow of soil water.

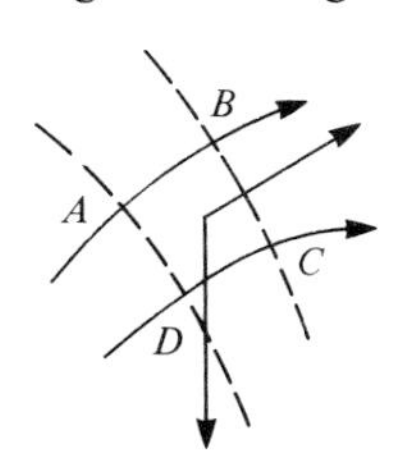

To sum up, the volume forces on an element of soil are:

$$(\vec{\gamma}_i + \gamma_w \vec{i})\,dV.$$

It will be noticed that the force due to the percolation, termed the seepage force, is independent of the permeability of the soil.[1] But the equilibrium of the forces could have been found in a completely different manner by considering the weight of water and soil within a certain volume, and then computing the pore pressure at a contour amongst the external reactions (fig. 10.2). The calculation consists therefore of using the total density and deducting at the boundary the effect of the uplift pressure u.

Obviously these two methods of calculation should be interchangeable, which can easily be verified graphically by drawing the forces diagram. This will moreover be demonstrated below in a simple example. Both of these methods of calculation have their advantages. The use of the total weight and of uplift pressure is still, however, the most convenient way of calculating the equilibrium of slopes when it is allied to the kinematic method of slip circles.

10.1.2 *Flow parallel to the surface of a slope*

Consider a slope inclined at α, subjected to a flow parallel to the free surface (fig. 10.3). It is a simple case of flow, more usual, moreover, in natural slopes than in those of earth dams. The hydraulic gradient is $\sin\alpha$. The first method of calculating the forces for the volume element gives

$$(\vec{\gamma}_i + \gamma_w \vec{i})\,dV,$$

whose components on the vertical and horizontal axes are:

$$Ox:\ \gamma_w \sin\alpha \cos\alpha \, dV$$

$$Oz:\ (\gamma_i + \gamma_w \sin^2\alpha)\,dV.$$

Let us now use the second method to calculate the forces in the element *ABCD* confined by two verticals and a line parallel to the slope at a depth h.

Fig. 10.2. Global equilibrium. Fig. 10.3. Flow parallel to the slope.

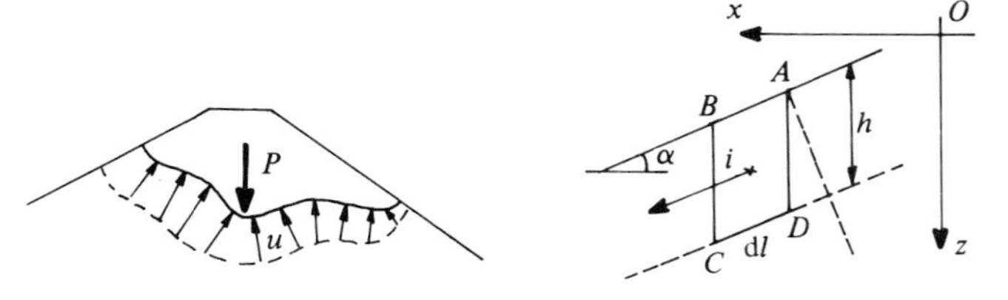

1 The case $\vec{\gamma}_i = -\gamma_w \vec{i}$ corresponds to a particular state of the soil known as a 'boiling' condition: there are then no longer any intergranular forces and the soil loses its resistance. This is the case for a sand which, subjected to a rising hydraulic gradient loses its bearing capacity and becomes a quicksand.

The pore pressures are equal on the segments *AD* and *BC*: the sum of their components on *Ox* and *Oz* is thus zero. The hydraulic pressure at depth h is: $\gamma_w h \cos^2\alpha$ and acts normal to the element of length dl inclined at an angle α. The weight of the volume element is γdV. Projecting on the axes we find:

$$Ox: \gamma_w h \cos^2\alpha \, dl \sin\alpha = \gamma_w \sin\alpha \cos\alpha \, dV$$

$$Oz: -\gamma_w h \cos^2\alpha \, dl \cos\alpha + \gamma dV = dV(\gamma - \gamma_w \cos^2\alpha)$$

$$= dV(\gamma - \gamma_w + \gamma_w \sin^2\alpha).$$

That the two methods are interchangeable is well proven since we know that $\gamma_i = \gamma - \gamma_w$ when the soils are saturated.

We can now use this flow model to find the greatest angle of stability of a slope of sand with an internal angle of friction φ and subjected to a flow parallel to the free surface.

Let us use the first method. The small element *ABCD* (fig. 10.3) is subjected to the following 3 forces (fig. 10.4):

1. its weight $\gamma_i h \, dl \cos\alpha$ (vertical),
2. the seepage force $\gamma_w \sin\alpha \, h \, dl \cos\alpha$ (parallel to the slope),
3. the reaction due to internal friction on the base *CD*; it is a force inclined at φ.

In equilibrium we find

$$\gamma_i \cos\alpha \tan\varphi = \gamma_i \sin\alpha + \gamma_w \sin\alpha,$$

or again

$$\tan\alpha = \frac{\gamma_i}{\gamma_i + \gamma_w} \tan\varphi.$$

In the majority of cases one can easily verify that

$$\frac{\gamma_i}{\gamma_i + \gamma_w} \approx \tfrac{1}{2},$$

and consequently, the limit slope is:

$$\tan\alpha \approx \tfrac{1}{2} \tan\varphi.$$

Fig. 10.4. Equilibrium of forces on the slope.

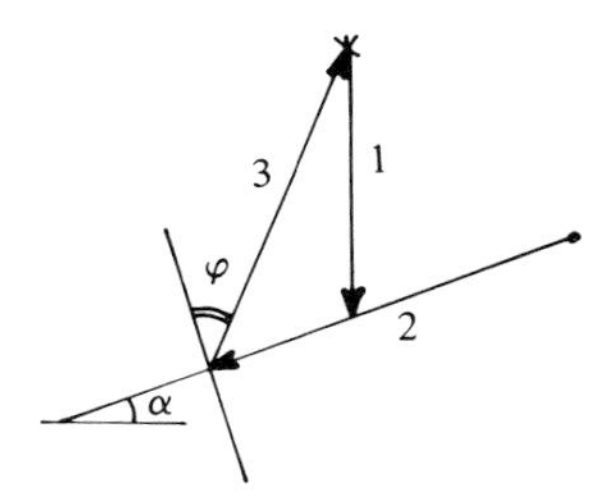

10.1.3 *Stability of the downstream face*

The stability of a downstream face is studied using the method of slip circles, along the slip circle the equipotentials of the seepage network allow one to determine the distribution of pore pressure. For equilibrium, the weight of all that lies above the slip line, that is the bulk density of the wet or dry soil above the saturation line, and the bulk density of the saturated soil below this line is taken into account. Obviously the presence of a seepage surface below the point defined by the intersection of the saturation line with the downstream face can be adverse to the local stability. The hydraulic flow at that point is almost parallel to the free surface, or more inclined, which is unfavourable, as can easily be seen using the preceding calculation and bringing $\vec{i}$ to the horizontal. It is absolutely necessary to ensure that the foot of the slope is stable by using a more permeable mass to constitute a local filtering surcharge (fig. 10.5), or even better to lower the water table by judiciously placing filters in the structure to avoid the appearance of a seepage surface. Remember that the transitory regime that allows the free surface of laminar flow to attain the theoretical status of permanent flow can last several years when soils are fairly impermeable. One or two seasons is not long enough therefore to assume that a downstream face is permanently stable if the seepage network is not perfectly controlled or monitored. Finally, the surface of the downstream face should be protected by a stony layer, or by grassing over when there is a risk of erosion by running water.

Fig. 10.5. Downstream slope of the Aït Aadel dam (Coyne and Bellier).

10.1.4 *Stability of the upstream face*

When the dam is in use the seepage force produces stabilising forces, which means that the upstream face can be steeper than the downstream face. However the upstream face is subjected to particularly dangerous forces when the reservoir is rapidly emptied. It is said to be in this condition of rapid drawdown when the water level is lowered at such a speed that the water that has seeped into the dam hardly has the time to move in a perceptible way in the earth mass. But the modification of the conditions at the boundary of the flow regime of the water immediately produces a modification in the orientation of the rates of percolation, that is to say of the seepage network and the pore pressure. This new network can easily be determined by electrical analogy, for example; fig. 10.6 represents an initial field of seepage and fig. 10.7 the same field disturbed by partial rapid drawdown. Using one or other of the calculations

Fig. 10.6. Infiltration network in an earth dam. Flow lines and equipotentials in a permanent flow.

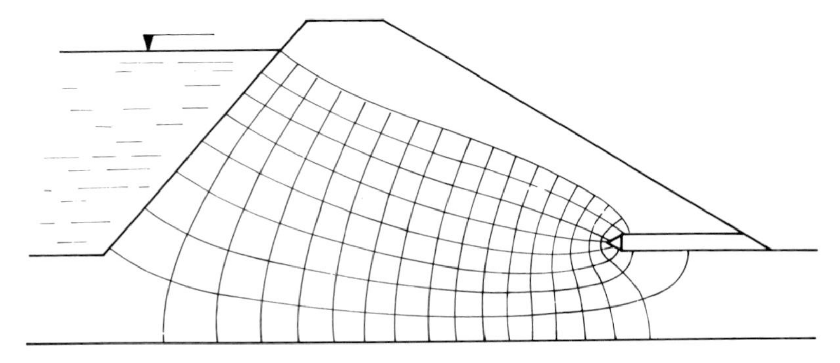

Fig. 10.7. Flow lines and equipotentials at the start of a transitory flow characterising partial drawdown.

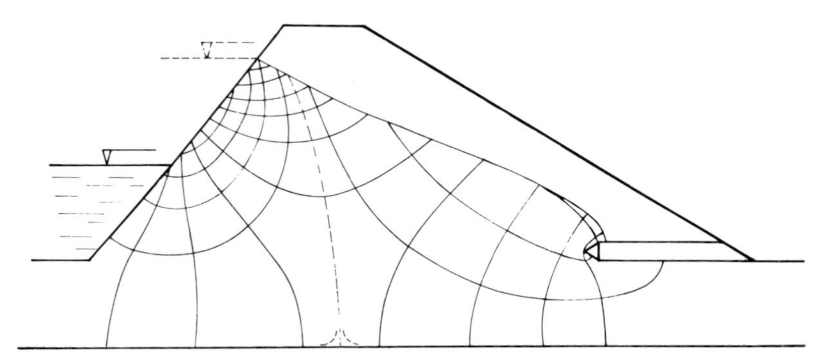

outlined above it can easily be seen that a total rapid drawdown is the least favourable case for the stability of the slope.

Here again, the calculation consists of studying the equilibrium of the least favourable slip circle taking into account the uplift pressures defined by a net of equipotentials and flow lines. To conclude, it should be noted that rapid drawdown is a relative phenomenon associated with the permeability of the soil. Thus, one could define intermediate drawdown as the seepage network evolving by a transitory regime, and being drawn down, as it were, by the gradual lowering of the level of the water at the upstream face. The limits of this phenomenon are revealed by considering the speed at which the water table in the body of the structure is lowered. As a first approximation one can say that water that flows in an almost vertical direction will have at most a hydraulic gradient equal to unity ($\mathrm{d}h = \mathrm{d}s$); an order of magnitude of the rate of the descent of the water table is thus given by its permeability. Therefore, one can say that if the permeability of the body of the construction is less than or equal to 10^{-9} m/s the drawdown is always rapid; if the permeability is more than 10^{-4} m/s problems with rapid drawdown rarely occur; but the problem does arise in the intermediate states, and, as an example, a permeability of 10^{-7} m/s and a lowering of the level of the water at a rate of 0.30 m/day should be considered as rapid drawdown. To estimate the rapidity of this effect with any precision, two phenomena need to be separated: firstly the drying out of the mass and secondly the consolidation of the earth which can, in some way, maintain the pore pressure. These two phenomena do not obviously occur on the same time scale.

Although rapid drawdown produces the severest effect it is a situation where the resulting accidents do not have any dramatic consequences: the reservoir is by definition empty at that moment, and the sole consequence of the rupture of the upstream face of the dike is to necessitate repair to the construction.

The banks of a canal can be considered as upstream faces. They are therefore in danger of rupture in conditions of rapid drawdown, and famous historical accidents have occurred on the Burgundy canal. The consequences of an accident in such circumstances are serious since the canal cannot be used until the repair has been carried out.

Finally, the surface of the upstream face should be protected from the wave wash action by laying down stone beddings or coatings before the water is let in.

10.1.5 *Stability of the foundation*

As for all structures placed on soil, it is necessary with earth dams to study their foundations. Once again the kinematic method of the slip circle is used, sometimes associated with the pore pressure, sometimes not. Generally, small constructions pose few problems for foundations. Even so the designer cannot afford to be neglectful, there are numerous traps, and a precise knowledge of the soil is indispensable. When working with large constructions the problems are of another sort: the sub-soil needs to be very carefully studied, as much from the point of view of stability as of settlement. Remember that information about the properties of natural soil is only obtained through samples taken from boring, and that it is likely to be a less uniform material than the compacted soil contained in the embankment of a dam – which is in effect an industrial product. This fact might very well lead one to use a larger safety factor for the foundations. Finally it should be noted that Jürgenson (1934) proposed an elastic design where the maximum shearing stress is compared at every point to the soil resistance. The local stability of the foundation can be rapidly established using this method.

In some projects it is necessary for the earth dam to be crossed by a canalisation (dewatering conduit, drainage gallery, etc.). Failure of such structures has caused many serious accidents. Indeed, a rigid inclusion in a deformable mass is subjected to stress concentrations; if the inclusion is linear it undergoes differential settlement which it cannot resist. If there is no possibility other than placing the canalisation beneath the dam, it is prudent to bury it in a trench hewn in the bedrock, which is subsequently concreted. If good soil is not accessible then sectional devices which have a certain amount of flexibility, or metallic culverts should be used (Habib and Luong, 1966).

10.2 Additional components: filters and waterproofing

The results of the stability studies carried out when a project is set up dictate the type and quantity of drainage and waterproofing that are placed in the earth dam to control the seepage network, modify the uplift pressures and lessen leak flow.

10.2.1 *Choosing the right size of filter*

As a filter is destined to control a seepage network, it is important that its efficiency remains constant – thus it should neither become damaged nor clogged. In a soil composed of grains of the same granular class it can be said that the average diameter of the holes is of the order of one-fifth of the average diameter of the grains; for a particle to be able to move in the interstices of such a soil its diameter must be half that of the hole. In this way the simple rule is formulated that a discontinuous filter does not serve as a barrier to grains that are more than ten times smaller. But it is very rare to find such materials in nature and sieving is a costly task. Most often then one should seek out acceptable natural products. It is therefore advisable to only use sands whose uniformity coefficient D_{60}/D_{10} is less than 2.

The following rules formulated by Terzaghi have been adopted for the relative grain size of filters and soils:

$$F_{15}/S_{85} < 4 \text{ or } 5$$
$$F_{15}/S_{15} > 4 \text{ or } 5,$$

where F_{15} indicates the diameter of the filter grains such that 15% in weight are smaller and S_{15} signifies the same for the soil (fig. 10.8).

When these two rules are observed one can be certain that the filter will neither become contaminated nor clogged. Sometimes it is necessary to use two different materials and to construct a layered filter which provides an effective barrier against the natural soil, whilst allowing good drainage at the same time – that is, it ensures the transit of a predicted infiltration flow across the structure, increased by a safety factor of about 10. Remember that the permeability of a granular soil is given by Hazen's relation with an accuracy of 50% which allows one to make an initial approximation of the required thickness of the filter.

Finally it should be noted that the preceding rules do not take into account the shape of the grains. Consequently, when faced with materials which are manifestly irregular (platelets, needles, etc.), whether these be in the soil or the filter, it is always easy and prudent to carry out direct tests to establish the stability or the absence of clogging of a material that might be used as a filter.

10.2.2 *Sealing curtains*

Sealing curtains have several roles; they are used in the following ways:

1. To lessen the discharge of infiltrations (fig. 10.9). In this case calculating the price of waterproofing almost always shows that the sealing should be total to be effective, and that the last few metres of sealing pay off most.
2. To elongate certain hydraulic passages and lower the gradients, and, in so doing, lower the risk of leakage and of piping. In earth dams, empirical rules are followed: the contact gradient is defined, for

Fig. 10.8. Rules for filters.

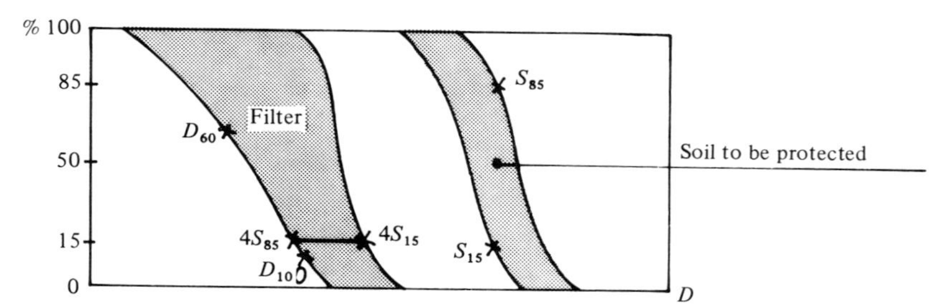

example as the relation between the heads of water and the length of contact between the natural soil and the structure; it should be less than certain values of the order of 0.2.[1]

3. Finally to remove uplift pressures whose size might otherwise endanger the stability of the structure. In this role they are often used in association with drains.

Fig. 10.9. Compacting the watertight core at the El Infiernillo dam (Mexico). Note the lateral filters separating the watertight core from the reloading masses. Also note the scouring on the bank, the lay-out of the different earth zones and the embedding of the watertight core at the rock. Height of the dam: 148 m.

1 Lane's rule takes into account above all the vertical flow paths Σh along the curtains corrected for the horizontal flow path L. It is written $i = H/[\Sigma h + (\frac{1}{3}L)]$. This limit gradient depends on the particle size distribution; it is of the order of 0.1 for silts, 0.2 for coarse sands and 0.4 for pebbles and gravels.

Grout curtains beneath foundations are connected to the water resistant components of dams – impermeable cores, cut-offs or blankets; the use of sheet piling has been practically given up nowadays in favour of curtains grouted into the soil, or cast *in situ* diaphragm walls. Various products are used to grout curtains depending on the particle size distribution of the soils being treated. In large-grained soils, that is soils with large voids, cement or cement mixed with clay is used, or again, if the holes to be blocked are really large, cement is mixed with sand and clay (Barbedette and Sabarly, 1953). In fine sands, soda silicate and the organic derivatives of silicates can be used, from which, with an

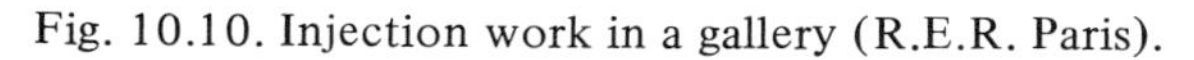
Fig. 10.10. Injection work in a gallery (R.E.R. Paris).

Fig. 10.11. Drilling and consolidation grouting in the Paris metro.

appropriate agent, silica gels are formed in the soil which are cheaper if less concentrated and less stiff. Organic resins are also used but they always lead to expensive solutions which can only be used in special cases. One should not assume that grouted curtains, or sheetpiling, create a cut-off that is as total as for a cast *in situ* diaphragm wall. In fact, what they do is to locally lower the permeability of the soil, and experience has shown that once this has been reduced 100 times the desired effect on the uplift pressures has been achieved (figs. 10.10 and 10.11). If the foundation of a dam is compressible, one should avoid using a rigid, undeformable diaphragm wall; in such a case one should build a diaphragm wall of clay mortar, which is more or less stabilised by cement so that the wall still retains a certain flexibility in the soil (Dupeuble and Habib, 1969).

10.3 Building earth dams

Site work inspection during earth dam construction will not be discussed here: control of the grain size distribution of the soil, of the water content at placing, of dry density reached, are all questions which have been looked at previously. The present survey will refer specifically to earth dam features – that is pore pressure during construction, settlements and infiltrations.

10.3.1 *Pore pressure during construction*

The embankment load causes the compacted soils to settle and the nearer the soil is to saturation, the greater the part of the surcharge borne by the pore water. Since the embankment is of large dimensions, the drainage is extremely slow, or at least it is not fast in relation to construction rate. The pore pressure is harmful in the sense that the shearing resistance does not increase with the surcharge since its expression is:

$$\sigma_t = c' + (n - u) \tan \varphi' .$$

To measure this pore pressure sensitive pressure gauges are placed in the embankment: vibrating wire manometers, electric strain gauge manometers, or controlled diaphragm manometers etc. surrounded by a porous body which supports the effective stress of the soil. Depending on the project, one should allow a rate of dissipation of pore pressure that corresponds to a certain percentage of the surcharge above the measuring instrument that should not be exceeded. Generally speaking it is advantageous to use materials for the embankment that are permeable enough for the pore pressures not to develop dangerously. Finally it should not be forgotten that placing measuring instruments in an embankment, whatever they might be, is a serious constraint in compacting.

10.3.2 *Settlements*

The compacting of earth has two goals: giving a better resistance and reducing the settlement of the structure. However, even with modern heavy compacting techniques, the lower parts of large structures can still settle again: and it is moreover, this settlement that causes pore pressures to develop.

The settlement does not have any serious consequences for homogeneous structures. In mixed or zoned dams the differences in settlement can bring about disorders in the area where they are connected together. It is thus desirable, if not to avoid them completely, at least to measure them. To do this one installs gauges at various points and monitors them by simple topography. The presence of these gauges and protection tubes is a significant constraint in compacting.

10.3.3 *Infiltrations*

When the reservoir is full, water filters through the earth dam, its foundations and its banks. It is generally a question of three-dimensional flow which is, however, easy to calculate by numerical means or electrical analogy since the seepage depends on a Laplace function.

When the structure is completed, it is important to ensure that the filtration flow is as was previously envisaged. It is thus necessary to follow the level of the water table or the pore pressure in the soil below the water table. In this way, the efficiency of the drains, the watertightness of the curtains, the existence of permeability heterogeneities in the foundation (privileged passages) or the possibility of disorders can be verified and remedied by further treatment. The instruments used for this vary from the simple piezometer, that is a tube in which one measures the water level, to much more refined instruments, similar to the pressure gauges used to measure pore pressure. It is in fact important that the instrument responds rapidly as it is evident that the variation of volume corresponding to the measurement of the variation of pressure by a pressure gauge disturbs the field of pore pressure after a certain time. This effect is even more sensitive the more watertight the soil. In loams with coefficient of permeability $k = 10^{-8}$ m/s, which are considered as impermeable soils for earth dams, ordinary piezometers are adequate if one takes the trouble to place a drainage cavity filled with sand at their base so that the response time corresponding to the filling of the tube is relatively short.

11

ROCK MECHANICS

Concrete dam foundations built on rock footings and underground constructions in rock present different problems from those which engineers have solved for earth works. The body of rules and techniques adapted to constructions in rock masses has, over the last twenty years been called rock mechanics. In fact, to resolve any given practical problem, the main task is finding out enough about the behaviour of a rock subjected to a force to be able to choose the most appropriate theory: elasticity, plasticity, strength of materials, soil mechanics, etc. At the present time rock mechanics is defined as the theoretical and applied science of the mechanical behaviour of rocky mediums and their physical environment; its uniqueness lies in the fissuration of rocks and in the discontinuous nature of natural masses.

11.1 Rock fissuration

11.1.1 *Mechanical properties of rocks*

The table below shows the resistance of the most common types of minerals to simple compression R_C, and to simple tensile stress R_T, and the elastic modulus E obtained by static loading.

It can be seen from this table that the word 'rocks' can mean many different things, obviously there is very little in common between granite and chalk: but even within the larger categories: sedimentary rocks (limestone, marls, gypsum, sandstone, puddingstone), metamorphic rocks (schists, marble, gneiss, quartzite),

	R_C (MPa)	R_T (MPa)	E (GPa)
Chalk	from 3 to 12	from 0.2 to 1.5	from 3 to 10
Limestone	7.5 to 45	0.5 to 4	8 to 30
Compact limestone	40 to 60	4 to 12	30 to 60
Sandstone	7.5 to 50	0.6 to 5	7.5 to 60
Schists	15 to 70	0 to 10	7.5 to 60
Granite	60 to 180	6 to 15	20 to 80
Quartzite	80 to 300	7 to 20	20 to 80

igneous rocks (granite, andesite, basalt, lava), there is no uniformity that would lead to a useful distinction of mechanical properties. An attempt at synthesis is thus indispensable if one wishes to discuss rocks as though they were one type of material.

A feature which all rocks share is their ability to be quite different while bearing the same name, the same chemical composition, and sometimes even the same geological source. The variations lie in their fissuration and in their structure. Natural minerals are, in fact, almost always fissured; the fissures are visible under the microscope in impregnated rocks as thin laminae and can even sometimes be seen by the naked eye; they are very flat voids whose thickness in relation to length is of the order 10^{-3} to 10^{-4}. The porosity corresponding to this volume of voids is very small, less than 1, or 0.5%, whereas there is often a much greater natural porosity, but associated with compact cavities. Fissures are situated between crystals, at the joints of the grains, or they are intracrystalline; their extents vary from one grain up to a hundred grains or even more.

11.1.2 *Mechanics of a fissure placing*

Consider a fissure within a fragile mineral subjected to an increasing force, a simple compression for instance, and examine its evolution in relation to the longitudinal stress (ϵ_1) versus strain curve and to the transverse deformation stress (ϵ_2) versus strain curve (fig. 11.1).

Right at the beginning of loading (phase 1) the fissure closes; the rock appears to be very compressible; the curve (σ_1, ϵ_1) is nonlinear and ϵ_2 is negligeable; Poisson's ratio (ν) is almost zero. When the stress increases (phase 2) the rims of the fissure slide over each other; the fissured material appears more deformable than the pure material; the transverse deformations are greater than in phase 1; the friction between opposing faces brings about some energy dissipation: the curves (σ_1, ϵ_1) and (σ_1, ϵ_2) are almost straight but a cycle of unloading gives rise to a hysteresis loop. There is, on the other hand, no hysteresis for volume variations, that is to say on the curve ($\sigma_1, \epsilon_1 + 2\epsilon_2$).

Above a certain threshold (3) the fissure yields and propagates progressively, usually in the direction of the largest stress. The matter becomes irreversibly damaged, which cannot be detected on the curve (σ_1, ϵ_1) (formation of small columns which do not alter the longitudinal deformation), meanwhile the transverse deformations increase, which is evident, and the curve (σ_1, ϵ_2) ceases to be linear. Poisson's ratio increases.

Poisson's ratio reaches a value close to or larger than 0.5 (and this bulking signifies that irreversible deformation has taken place) when phase 4 begins and the different fissures join up. At the moment the curve (σ_1, ϵ_1) in turn ceases to be linear, and permanent deformation appears when unloading. When the threshold (5) is reached rupture takes place instantly; between (4) and (5) the phenomena of delayed rupture occurs. Experience shows that the length of life of the sample is longer the closer the load is to (4) which is sometimes called the ultimate load (infinite length of life).

This description is schematic and depends on the orientation of the fissure. If the fissure plane is close to the normal to the compressive force phase 1 will be quicker and phase (3) slower or even non-existent. In reality there is a statistical aspect to the process since the orientation of the fissures is random.

The preceding analysis shows that at the moment of loading a rock, information about deformation in the direction perpendicular to that of the major force is particularly important: very valuable information can be obtained from the corresponding diagram about elasticity, fatigue, damage and the rupture of rocks. Finally, this description is that of the behaviour of a medium containing fissures of finite extension. But it could easily be transposed to the behaviour of the fractures of a rock mass (schistosity, diaclases, stratification).

Fig. 11.1. Stress/strain curves for a fissured medium.

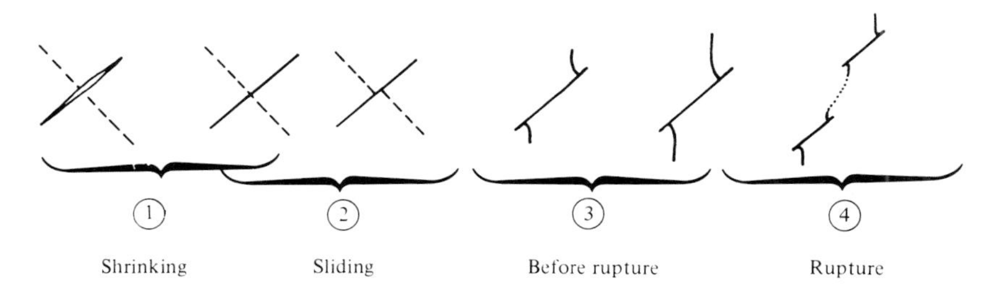

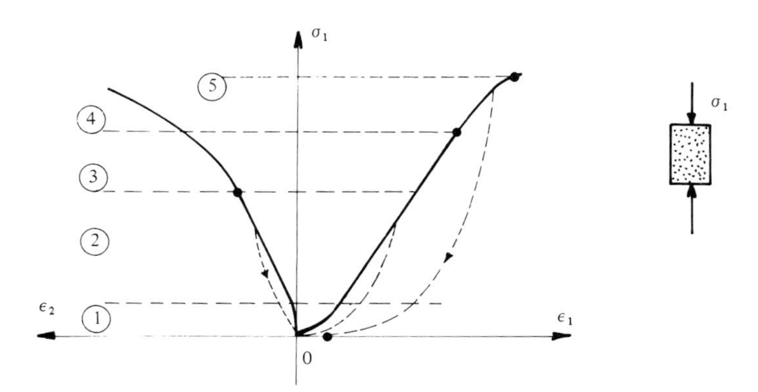

11.1.3 *Effects of rock fissuration*

We have just seen the influence of the fissuration of rocks on the stress/strain curve. To understand the concept of rock fissuration is to understand the whole of rock behaviour. It suffices to point out the following principal phenomena:

The elastic modulus of the rock is lower than the elastic modulus of the constituent minerals.

The anisotropy of a rock can be much greater than that of the constituent mineral crystals (the fissures only need to be oriented; e.g. schists).

The permeability of a rock in one direction depends on the stresses (closing of fissures or opening of new fissures).

The speed of propagation of sound waves in one direction depends on the stresses (*idem*: a fissure obstructs the propagation of sound).

The tensile strength is more dispersed than the compressive strength (fissures exercise more influence over the tensile strength than over compressive strength).

Tensile strength in an imposed plane for example in a line load test[(1)] is greater than the tensile strength of a certain volume (simple tensile test).

Simple compressive strength appears to present a scale effect: the larger the sample the lower the compressive strength (greater probability of having a major fissure in a large sample).

If the mean stress increases, the rock loses its brittle nature and becomes ductile (under a high hydrostatic pressure the fissures close, the friction between the walls reestablishes the apparent mechanical continuity and it is the crystalline edifice which bears the stresses). This rule is very general but the mean pressure at which the ductile transition takes place depends on the base mineral and its hardness: 5 to 10 MPa for gypsum or saturated chalk, 100 to 200 MPa for compact limestone, perhaps 10 GPa for quartz.

11.2 Fissuration of rock masses

11.2.1 *Mechanical properties of rock masses*

Fig. 11.2 shows the fissured face of a rocky mass, in contrast to fig. 11.3 which shows the appearance brought about by mechanisms of dissolution.

1 The line load test, or Brazilian test, is a rather specific but very convenient test. A force P increasing to failure is applied in two opposite generating lines of a cylindrical sample of diameter D and length L. The tensile stress perpendicular to the plane of the two generating lines is constant and equal to $R_T = 2P/\pi DL$.

Fig. 11.2. Fissured facies. Three families of fracture planes are visible: inclined to the right and white with snow, inclined to the left and grey (grass), parallel to the plane of the photograph and black. Horizontal distance: about 600 m.

Fig. 11.3. Facies showing dissolution. The ravines of this gypsum mountain are due to dissolution by rain and mountain snows.

The properties of rock masses are associated with natural fissuration, and all that has been said about samples applies to these masses – that is that deformability of a rock mass is essentially that of its fissures, and that the resistance of a rock mass is linked to that of its joints. In the same way again the circulation of water in the mass is determined by the fissures, and the fissures are obstacles to the propagation of sound. A mass is therefore compact when it is watertight and when the speed of sound waves is very high,[1] and vice versa. But the fissures of a rock mass are of considerable extent – they are macrofractures, open or closed, joints with or without filling materials. Geology and tectonics play an important role in the preferential orientation of certain systems of joints or natural fractures (stratification, schistosity, diaclases, faults, slip lines) but frequently similarly orientated microfissuration on the crystal scale occurs to correspond with the discontinuous fissuration on the scale of the rock mass.

When a rock mass ruptures, unit blocks dislocate around their joints, accompanied by a significant increase in volume. A natural fissure sometimes elongates, but it is very rare for a rupture to cross a rock that is initially sound. Finally, the way in which stresses are distributed under the effect of a force is very much influenced by the presence of joints. Fig. 11.4 shows the isochromatics of a punching test on a two-dimensional mass using photoelasticity. It is the classic Boussinesq–Flamant problem. Fig. 11.5 shows the isochromatics

Fig. 11.4. Isochromatics of a punch on a semi-infinite mass.

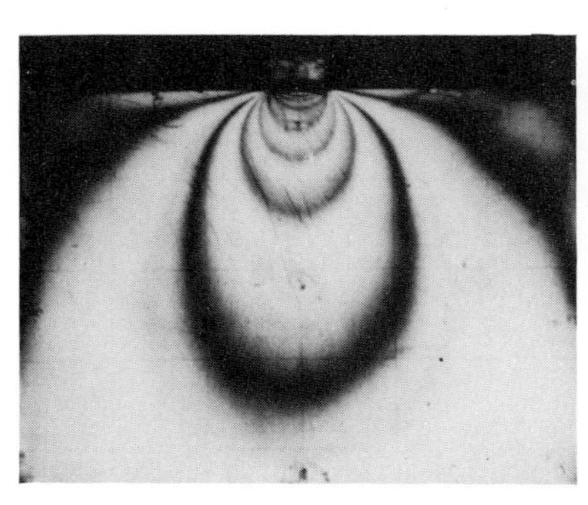

Fig. 11.5. Isochromatics of a punch on a stratified, semi-infinite medium.

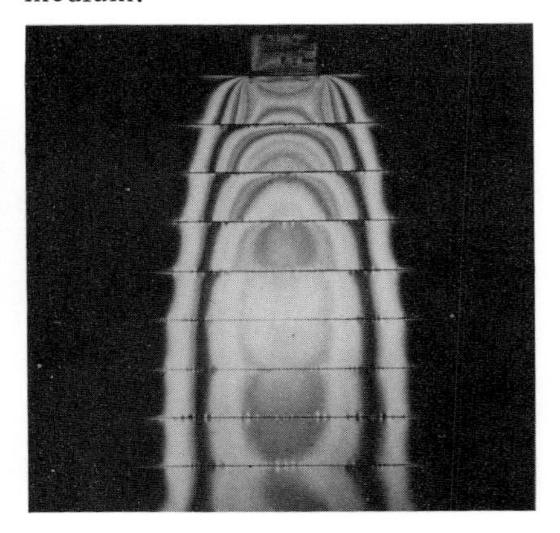

1 In the years before the accident of 9 October 1963 which precipitated 300 million cubic metres of rocks from Mount Toc into the dam reservoir at Vaiont (Italy), significant slipping movements had taken place in the rock which brought about fissuration. Measurements of the speed of propagation of sound waves showed a variation of velocity of longitudinal waves from 5500 m/s to 1900 m/s before the collapse.

in a stratified medium for the same problem (Maury, 1970). The stresses plunge downward and in the second case the zone affected by the large loads is narrower, correlatively and excluding joints the settlement is much larger under the punch in a stratified medium. This result can easily be explained. In a medium broken up by joints which can only support small shearings, tensile stress is not transmitted through the mass. Tensile stress can only be found in the material on the outer edge of a slab that is bending, and the deformation of the whole results from the superposition of all the elemental bending. Thus the transmission of stresses at depth which appears to be totally natural when the stratification is nearly vertical, also occurs for a horizontal stratification, which is not *a priori* evident.

All these phenomena show that it is essential to know the geometrical and resistive properties of the joints.

11.2.2 *Investigation of a rock mass*

Rock masses do not lend themselves easily to having intact samples lifted. Simply viewing the rock mass from inside a gallery reveals that it is divided by several systems of fractures. A small sample is not therefore representative of the whole, and if knowledge of the mechanical properties of the matter itself, i.e. of the mineral, is a necessary first step, it is not enough unless accompanied by a study of the body of the rock mass and the properties of the joints.

The first task then is to proceed with the examination of the fissures and to study the geometrical distribution of families of fractures. There are often three families, but their orientation can vary from one point to another. A convenient representation of fissure networks of a site can be gained by stereographically projecting the trace of the normals on a sphere (fig. 11.6). Unfortunately, such a description is always incomplete since the only information known about that fissuration is its path at the surface, allied with a few rare observations in the interior by means of galleries or bore-holes. To penetrate to the interior of a rock mass and gain a more global picture, geophysical methods are used.

In a continuous medium the theory of wave propagation gives the following classic equations for the speeds C_1 of the longitudinal waves and C_2 for the transverse waves:

$$C_1 = \left[\frac{E}{\rho}\left(\frac{1-\nu}{(1+\nu)(1-2\nu)}\right)\right]^{1/2},$$

$$C_2 = \left[\frac{E}{\rho}\left(\frac{1}{2(1+\nu)}\right)\right]^{1/2}.$$

If these equations are applied to rocks and rock masses i.e. a fissured medium, the terms E and ν, sometimes qualified as dynamic, can be determined and are different from the static values. In particular the dynamic modulus E is higher than the static modulus E and all the more so when the rock is more fissured. Cases have even been cited where the ratio of the first to the second is as high as 10. This example shows that applying the theory of elasticity to rocks is often

Fig. 11.6. Stereographic representation of the fissuration of a site and of the position of the two slopes of a valley.

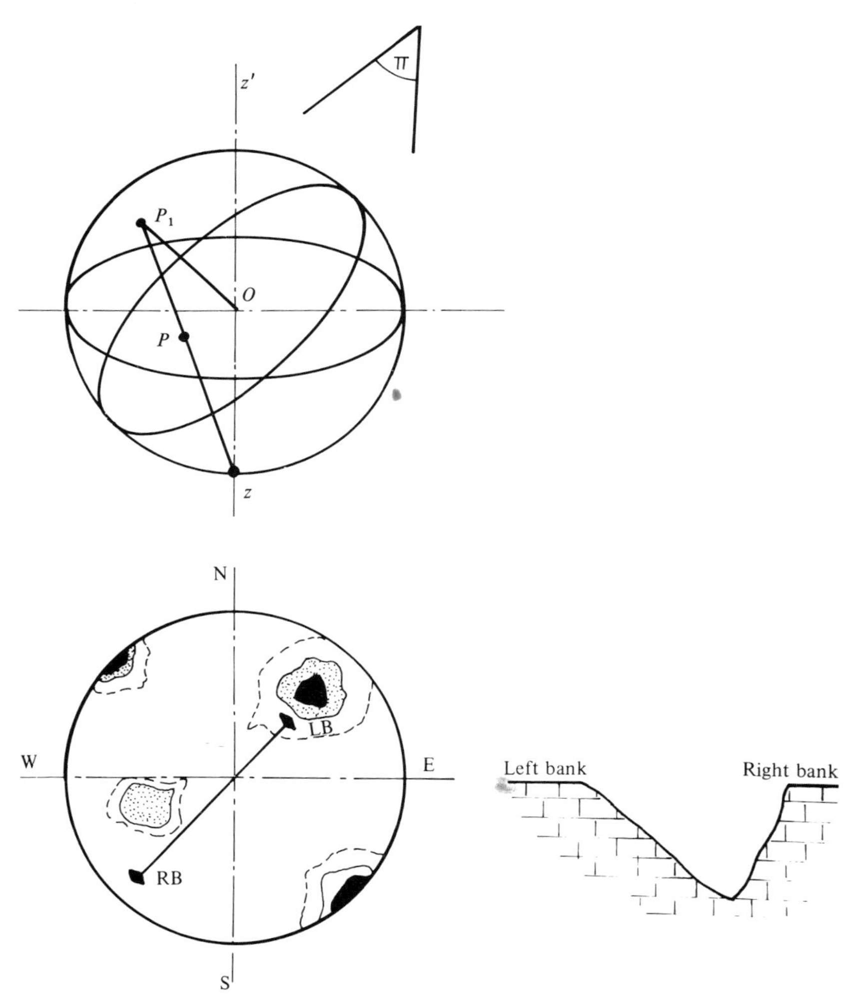

only a rudimentary approximation. Even so, seismic methods are at present the most convenient procedures for examination of rock masses in order to evaluate the homogeneity of a large volume. In addition to the large irregularities such as faults, they show the extent of the decompression zones, which are identified by slower speeds. Even if interpretation is difficult in mechanical terms, the identification of zones where the wave speeds are different is very useful information particularly if it relates to fissuration at the site.

11.2.3 *In situ measurements of mechanical properties*

Discovering that mechanical measurements made on small samples had little significance, it occurred to engineers to transport the laboratory to the rock site and to set up large forces which would affect a considerable volume of rock in order to bring a significant number of fissures into play.

Measurements of deformability are made in order to define Young's modulus. The punching method can be cited where a circular area of the order of 30 cm is applied to the surface of a semi-infinite mass. The displacement for a punch of radius R is

$$w = \frac{P}{2R}\,\frac{1-\nu^2}{E}$$

if the punch is rigid. The load P is of the order of 100 tonnes. One can also cite the method whereby a dilatometer is expanded in a cylindrical cavity; if p is the pressure applied, the displacement is

$$\frac{\Delta r}{r} = p\left(\frac{1+\nu}{E}\right),$$

where r is the radius of the cavity, from which one can find E by giving a reasonable value to ν.

When the geometrical definition of the extent of the fissured zones and the orientation of the fissures has been found, one carries out measurements of the rupture of joints, or of the most dangerous family of joints. Tests carried out on

Fig. 11.7. Dilatancy of a joint.

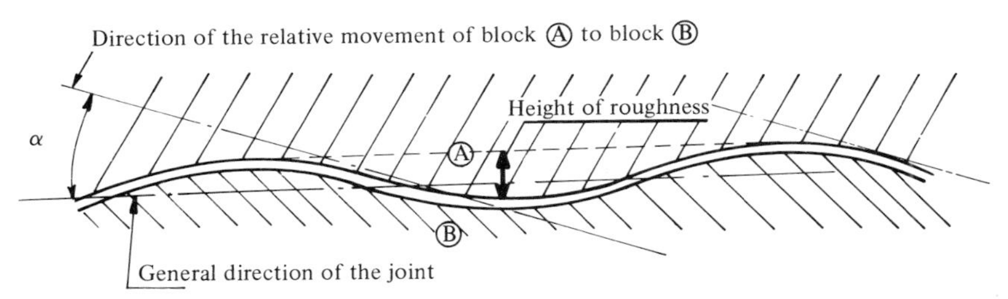

family is once again possible, but only in one direction. Therefore block *C* in fig. 11.8 can only move in a downward direction.
several square metres of the surface of joints are common. Experience shows that on the one hand, the equilibrium conditions for joint rupture are analogous to those of Coulomb, i.e. $\sigma_t = c + \sigma_n \tan \varphi$, and on the other hand there does not seem to be any scale effect for the joint resistance other than that linked with the roughness of the surfaces present, which we shall now discuss.

The roughness of the joint is a doubly interesting concept. In fact, when it is defined in the broad sense of fig. 11.7 and one can estimate an angle α of inclination of the roughness in relation to the general direction of the fissure, it is clear that the conditions for rupture should be modified to: $\sigma_t = c + \sigma_n \tan (\varphi + \alpha)$. But on the other hand, fig. 11.7 determines the dilatancy of the joint, that is to say the bulging of the rock; it is clear that the same angle α can be associated with different degrees of roughness, in other words with dilatancies that are of different magnitude, and hence with very different stabilities.

11.2.4 *Kinematics of sliding*

If it is only in exceptional circumstances that a rupture takes place from the interior of a rock mass then that must mean that the failure of a mass is produced by rotation and movement of the unit blocks (e.g. collapse under hydraulic pressure acting in the fissures) or by translation, that is to say sliding along the joints. It is the second mechanism which is the most important, and stability calculations (Londe, 1965) are made principally by studying the equilibrium of a unit block that is able to slide over each of its supporting faces, or over two of them.

The kinematic aspect must be stressed, as must examination of every possibility for movement. Fig. 11.8 shows two families of joints in a mass, and it is assumed that a system of forces brings about shearing. Depending upon the orientation of the principle directions and the equilibrium conditions for each family of joints, sliding can occur, for example in the horizontal direction. The corresponding movement prevents any further action of the second family of joints: block *A* having moved to the right, block *C* can no longer slide upwards. If the system of forces now changes, the movement of the joints of the second

Fig. 11.8. Displacements of blocks defined by two families of joints.

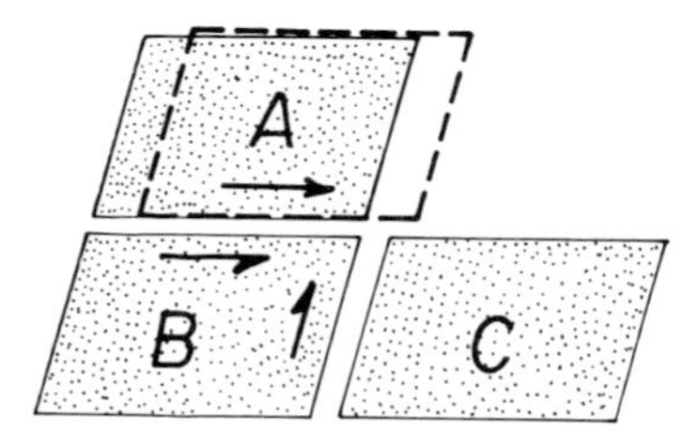

These considerations can be of the utmost importance in practice, which we will illustrate by a simplified real case. Fig. 11.9 represents a dam site where the stratification is almost a flat surface and inclined at 35° to the horizontal; the strata form a 'stacking up' effect of slabs which become unstable if the passive resistance at the foot of the stratification is removed, which happens when the dam foundations are excavated. If one looks at the stratification plane (fig. 11.10), one discovers two families of joints (I) and (II) (fig. 11.11). By close examination one can verify that the family of joints (I) has already acted, which has given the family (II) a bayonet-like form. This description and the associated kinematics gives totally different situations for the right and left banks (fig. 11.11). When excavating on the right bank only a limited number of blocks threaten to fall, which can always be held back by anchorage. On the left bank on the other hand, the problem is very different; the size of the zone likely to slide is very much larger and the precautions needed to hold them back are incomparably larger.

11.2.5 *Weathering of rocks*

If it is essential to know what the resistance of a rock is at the time of building constructions on it, it is important to ascertain the long term stability of the natural materials. Now, contrary to what one might believe, rocks evolve

Fig. 11.9. Inclined stratification.

Fig. 11.10. Stratification plane.

Fig. 11.11. Model of the stratification plane seen from the downstream face.

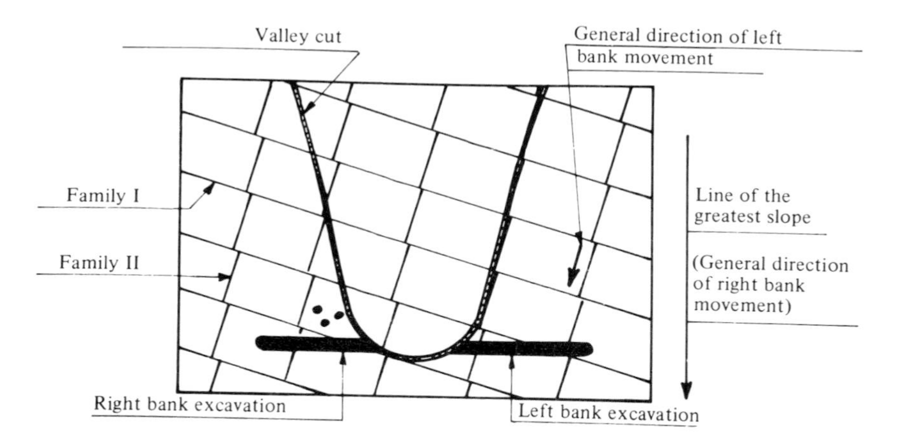

rapidly and are capable of losing their resistance in very short periods – periods which are all the shorter when explosives have opened up or increased their natural fissuration.

Weathering of rocks is due in general to the action of a fluid, most often water. When a fluid penetrates into a rock and makes contact with certain constituents, chemical changes take place. The loss of resistance of a rock by weathering is associated either with the phenomenon of dissolution or that of swelling which can bring about disorders in the mineral structure. Implicit in the state of weathering then, is partly the specific weathering of a mineral (Farran and Thenoz, 1965), and partly the possibility of a fluid accessing the rock through micro- or macrofissuration. Once again fissuration appears to be the major characteristic of rock material.

Measuring the porosity of rocks is not likely to furnish useful information, since the volume of the fissures is insignificant. Measuring the permeability provides the most interesting results. Thus, for a given type of rock, the scale of permeability is in proportion to the scale of weathering. But this is not the only result that can be obtained in this manner. Analysing a liquid which percolates through a rock permits one to detect the elements that are being carried away. A diminution of the flow during a test signifies that displacements are occurring in the circulation networks and denotes, generally, a rock sensitive to weathering.

Finally, if one modifies the effective stresses in a rock, one can establish the variations of permeability in relation to the opening or closing of the fissures. It is the same if the temperature of the fluid varies (transitory seasonal regime) and Jouanna and Rayneau (1971) have pointed out that the effect of thermal expansion of the rock is capable of lessening the width of the fissures and consequently changing the network of hydraulic equipotentials.

11.3 Foundations on rock masses

11.3.1 *The problem of punching*

Consider a small punch such as that used for trepanning in drilling. The resistance to punching is given practically by the cohesive term of the general formula for bearing capacity, that is to say $c\,N_c(\varphi)$ (see section 7.3.1) where c and φ are the coefficients of the intrinsic curve for the rock and N_c is the cohesion factor. This intrinsic curve is determined in the laboratory with apparatus that is, in principle, identical to that used for soil mechanics, but very much more powerful. It has a parabolic form; c and φ are then the mean parameters of a Coulomb line adjusted to best represent a curve of parabolic appearance. One sees immediately that the punching pressure easily reaches several GPa.

Consider now a much larger punch such as the pressure exerted by the foundation of a construction. The bearing capacity formula is the same, but the values of c and φ may be different because of the scale effect. In certain cases c and φ can be much smaller, especially the cohesion which can reach zero. Furthermore, if the foundation is broad, the deformation of the footings at rupture are large since they express themselves as percentages of the breadth. They are thus incompatible with the majority of deformations admissible in construction work. The calculation for the foundation thus amounts to establishing the statics of the different loaded elements constituting the rock mass, and to finding an elastic displacement of the foundation that is compatible with the deformations of the structure.

Let us note finally that rocks, like soils, sometimes show delayed deformations. Deformations from consolidation are unusual and then only small. Deformations from creep can form a significant fraction of the instantaneous deformation, notably in chalk, but they rarely affect civil engineering on the surface.

11.3.2 *Foundations of dam footings*

There is only a small quantity of water in a rock mass since the porosity is small. One might wonder whether pore pressure plays as important a role as in soil mechanics. A simple analysis of the problem in a medium where the fissures are continuous leads immediately to the relation $\sigma = \bar{\sigma} + u$ for the total stresses σ and the effective stresses $\bar{\sigma}$ and this leads, again, to Hvorslev's equation (section 2.1.1):

$$\sigma_t = c' + (\sigma_n - u) \tan \varphi'.$$

Experience shows that for discontinuous fissuration, like that found in the rock itself, between the diaclases, when the fissures are flat enough and distributed at random by the crystallisation, the law of effective stresses is also valid.

It has been noted above that the permeability of rocks is affected by the stresses. The result of this is for the permeability of a fissured mass to decrease with depth because of the geostatic pressure. The flow in the fissures follows on average the laws of proportion to the hydraulic gradient (even if the movement is locally non-laminar), but the permeability of the first 10 m is generally greater.

Likewise, under the effects of a surcharge of a dam footing the permeability can be reduced locally and the distribution of the uplift pressures is completely modified from that determined for a homogenous mass of earth. This phenomenon is essential in calculating the stability of rock slopes on the downstream face of a dam, and since precise calculations are lacking, one must place piezometers in a fissured mass to check the pore pressure. An investigation into the position of the watersheds forms an integral part of the preliminary study of a site before the construction takes place.

11.3.3 *Other problems in rock mechanics*

Rock mechanics is not solely used in public works. It is encountered in numerous other industries. But the size of the applied stresses is variable and determined by the demands of the designer. Civil engineering works, whether constructions or quarries, are almost always carried out near to the soil surface: the stresses transmitted to the rock are several MPa. Mines and tunnels are constructed at depths varying from 100 to 1000 m and the stresses are of the order of 2 to 50 MPa. Finally, oil drillings can descend as far as 4000 m and encounter rocks which can stand 100 MPa. The length of time during which loads are applied is also variable: several weeks while a mine is being built, a seasonal cycle for structures retaining water, a few scores of years for the foundations of construction works, but a few milliseconds for the stresses brought about by an explosion when works are excavated using explosives.

11.4 Improvement of the mechanical qualities of rock masses

In general, in mechanics, if the material does not have the necessary qualities, all one has to do is to change it. For instance, special steels can be used if ordinary steel is not good enough. In civil engineering and especially in rock mechanics, one is forced to work with the existing medium and adapt to it as one finds it. And, since the forces due to the weight of the rock masses are sometimes extremely large, in the majority of cases it is hardly possible to combat them by building resistant structures. One can only act economically on fissuration, and even then the methods available are few and far between: groutings, drainage, rock bolts and anchorages.

11.4.1 *Grouting*

Grouting has a double role. Most often it is used to construct grout curtains that aim to displace the uplift pressures, and consequently modify the orientation of the hydraulic forces in the interior of the mass to improve the general stability. Under a concrete dam, for instance, the grout curtain in the rock would be tilted towards the upstream face. The concept of watertightness is associated with a diminution of seepages in a water-retaining structure, and with a lessening of erosion or weathering of the footings of masses. Grouting can also, though, have a mechanical role in the filling and closing of fissures (which increases the elastic modulus) or blocks may be sealed together which ensures that they are, to a certain extent, consolidated.

In rock masses, grout curtains are always made by boreholes and injections, at a fairly close distance to one another. One tries to use high pressures so that the grout penetrates into the mass but without however causing any disorder in the body of the structure. Generally, injected products are made from pure cement grout or mortars, or mixtures of cement and clay. Washing out the fissure by injecting water and bicarbonate of soda under pressure is a highly uncertain operation, but it is desirable to seek a firm adhesion between the mortar and the rock. It has been established that clay-cement mixtures have a better adhesion in the fissures when the mineralogical nature of the clay is similar to that of the filling material of the fissures. Cement injection can fill the voids around a subterranean structure and improve the behaviour of the rock mass. These last years have seen the use of resin injections (epoxy or polyester) in rock masses where before their use had been restricted to soils.

Finally, the efficiency of the grout curtains can be checked by placing piezometers in the downstream face of the curtain, or perhaps also in the upstream face, and by measuring the flow of the rising waters returning to the valley.

11.4.2 *Drains*

Drains are used to dispel the uplift pressures in places where their presence is particularly dangerous. The density of drains is determined by the fissuration of the mass. They should be cut through all fissures capable of allowing water circulation; the lowering of the water table which they cause determines their effectiveness and radius of action, and this can be checked, as with grout curtains, from observations made by piezometers previously placed in the mass at a certain distance from the drains. The diagram in fig. 11.6 defines the direction of the boreholes and allows one to install the drains in the correct position. Thus the fissures or fissure network where uplift pressures dangerous to the work might develop (as shown by calculation) can be drained. Lastly one must avoid grouting once the drains have been installed, or one runs the risk of blocking them.

11.4.3 *Rock bolts and anchorages*

Stresses that cannot be tolerated at the surface can be transferred to greater depths using anchorages. They give the fissured body a cohesion which one can try to estimate using the principle of corresponding states.

Nowadays it is possible to produce prestressed anchorages of several hundred tonnes by grouting metallic cables or bars sealed in mortar into the boreholds. The active part of the tendons is protected from corrosion by a sheath which is usually grouted. It is advisable to monitor the tension of the cables, particularly when there are blastings nearby which creates the danger of rupturing the fixed anchor length.

Support by short anchorages (2 to 4 m) is called rock bolting; it is frequently used in galleries with forces of 4 to 10 tonnes. As they have now been in use for about ten years we have a statistical idea of the efficiency of different systems used (Tincelin *et al.*, 1971). Rock bolts anchored at the foot, for example by wedge-shaped devices, and pretensioned with a dynamometric key at the head, but free in the cavity, do not give such good results as those gained with bars grouted along their length with resin and obviously not prestressed. The deflection of the roofs of mines have been on average 25% less with grouted bolts than with anchored bolts. With fixing, smooth steels or crenellated bars give similar results. Blasting can cause a loss of tension of up to 75% in anchorages.

12

EQUILIBRIUM OF UNDERGROUND STRUCTURES

Rigorous study of the equilibrium mechanics of a subterranean gallery of any shape and alignment has not yet yielded solutions, and is far from doing so. However, various simplifications exist giving explicit solutions that provide an approximation that is adequate for current technical problems. In particular, the cross-section of a circular gallery is often used as an acceptable picture of the actual structure. From the point of view of calculations this form has the advantage of axial symmetry and the problem can be reduced to one plane.

12.1 The mechanism of collapse of an underground structure

An underground structure can either be stable under the weight of the earth that lies on top of it or it can collapse and cave in. As the weight of earth is determined by the depth, loading immediately occurs at the moment of boring. There is no progressive application of stress as there is when a dam is being filled with water, or during the loading trials prior to opening a bridge. The concept of a linear relation between the stresses and deformation to define elasticity becomes difficult to perceive since the stresses are instantly constant. The idea of an elastic domain still remains an essential one and can easily be reestablished by imagining an initial situation where the density of the earth is zero and the boring carried out, then by increasing the density from zero to its real value and following the progression of the deformations inside the mass.

It should be noted moreover that the vertical force due to the weight of earth is not the only force, there is often a horizontal reaction: earth pressure at rest in soils, orogenic stress in rock masses. Generally little or nothing is known about this stress.

Another paradoxical aspect of the equilibrium of subterranean structures lies in the fact that except where the methods of execution are very unusual, one can say that a gallery always goes through a certain period of time when it supports itself; this could for example be the few months that elapse between the boring and the concreting; or perhaps the few minutes between the installation of the first part of the arches and the last part of the banking. The pressures which appear subsequently behind the lining are thus due to delayed deformations, that is to say to the rheological behaviour of the material. In soils this adaptation can take several months and in rocks several years. The value that these stresses can attain in the latter instance shows the importance of the viscoplasticity of rocks, which is not evident *a priori*. In this way the railway tunnel at Mont Fréjus in the Alps, between Modane and Bardonecchia, is over 100 years old and situated in sound rocks (calcschistes). The normal stresses of the rock mass acting on the linings of the masonry are of the order of several tens of MPa (Bernède *et al.*, 1968), which is on the one hand far from being negligeable for a thin arch of some ten metres span, but does not on the other hand bear any relation at all to the thickness of the overlap. It is clear that such stresses did not exist when the linings were put in place, and rock creep must have taken place over several years before reaching the present equilibrium.

After all, the stresses which act on a lining are very much more conditioned by the shape of the gallery, the mode of execution and the rheological behaviour of the soil than by the thickness of the overlap.

Before studying the plastic domain, we shall consider equations which verify whether the material in which an underground structure is bored is in an elastic state.

12.2 Elastic analysis of a circular gallery

The theory of elasticity completely resolves the problem of a circular cavity in one plane. For other forms it is always possible to use experimental methods, especially photoelasticity.

Consider a circular cavity in an undefined mass subjected to a uniform field Q of compression. Using the notation of fig. 12.1, classical theory (Timoshenko and Goodier, 1961) gives the following values for the stresses:

$$\sigma_r = \tfrac{1}{2}Q(1-a^2/r^2)+\tfrac{1}{2}Q[1+3(a^4/r^4)-4(a^2/r^2)]\cos 2\omega,$$

$$\sigma_\omega = \tfrac{1}{2}Q(1+a^2/r^2)-\tfrac{1}{2}Q(1+3\,a^4/r^4)\cos 2\omega,$$

$$\tau_{r\omega} = \tfrac{1}{2}Q[1-3(a^4/r^4)+2(a^2/r^2)]\sin 2\omega.$$

When the plate is subjected to two different stresses, P and Q, the value of the stresses at all points can be obtained using the principle of superimposition.

Fig. 12.1. Circular gallery in elastic theory: notation.

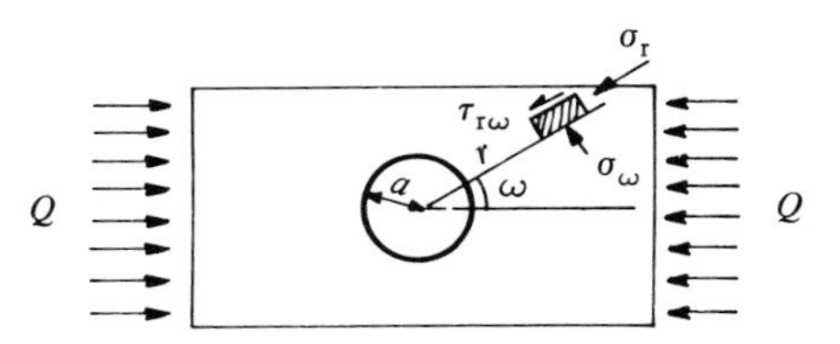

Fig. 12.2 is a geometrical representation of the results obtained for σ_ω on two orthogonal axes placed in the principal directions. P is the vertical stress and Q the horizontal stress. Different cases are shown. If $P = Q$ it can be seen that the stress tangential to the wall of the gallery is equal to $2P$. If $Q = 0$ then at the keystone there is a tensile stress equal to $-P$ and on the side walls a compressive stress equal to $3P$. If $P = 3Q$ the tangential compression is equal to $8Q$ at the side walls but zero at the keystone.

Let us suppose now that a pressure p is acting at the interior of the gallery. Lamé's solution for thick tubes gives:

$$\sigma_r = p\,a^2/r^2\,,$$

$$\sigma_\omega = -p\,a^2/r^2\,,$$

Fig. 12.2. Distribution of σ_ω around the circular gallery for different loads (elasticity).

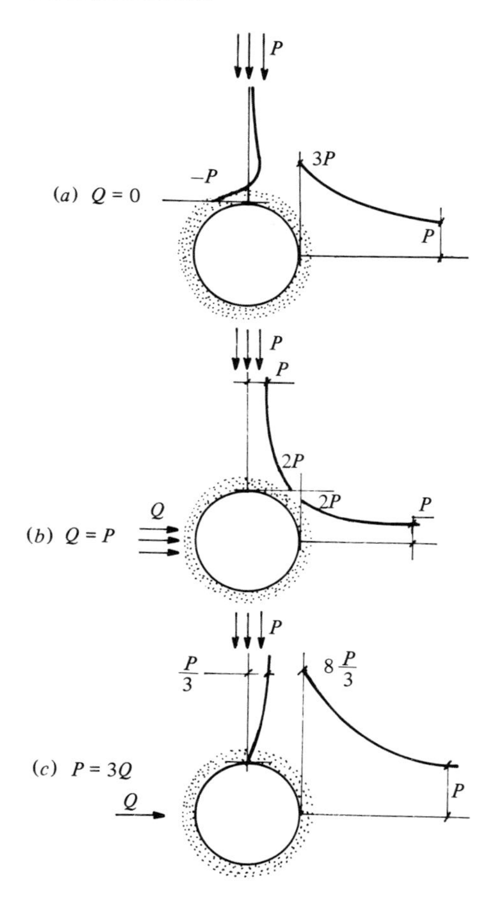

$$\sigma_z = 0,$$

where σ_z is the vertical stress, which is graphically represented in fig. 12.3.

Superimposing on the preceding equilibria we obtain the solution to the major current problems. Fig. 12.4 gives for example the case $Q = 1.4P$ with $p = 1.6P$ showing the appearance at the keystone of a tangential compression of $1.6P$ and zero stresses at the side walls.

The case of zero tangential stress which has already been cited twice in the preceding examples is important since it represents an interesting practical limit, in fact, from that moment the risk of rupture is great. Soil, for example, no longer has any tensile strength, nor will a rock mass that has been more or less disrupted by blasting (or by natural fissuration). When such a situation arises at the keystone one can be just about certain that the roof will begin to cave in.

The present analysis neglects the stress perpendicular to the cross-section of the tunnel; in certain cases this becomes inadequate. For example, in a vertical circular shaft the weight of the earth intervenes and the stresses around the shaft depend on the depth z. It could be that σ_z is not the intermediate principal stress but the major principal stress. It is thus necessary to know all the stresses to estimate when one has gone beyond the elastic state. Shown below are Kirsch's equations where σ' and τ' are the variations of the stresses in the hypothesis of plane deformation between the initial and final states after a gallery has been excavated. P and Q are the vertical and horizontal pressures extending to infinity and ν is Poisson's ratio:

$$\sigma'_r = [-\tfrac{1}{2}(P+Q) + \tfrac{1}{2}(P-Q)(4-3a^2/r^2)\cos 2\omega]\,(a^2/r^2),$$

$$\sigma'_\omega = [\tfrac{1}{2}(P+Q) + \tfrac{1}{2}(P-Q)(3a^2/r^2)\cos 2\omega]\,(a^2/r^2),$$

$$\tau'_{r\omega} = \tfrac{1}{2}(P-Q)(2-3a^2/r^2)(a^2/r^2)\sin 2\omega,$$

$$\sigma'_z = \nu(\sigma'_r + \sigma'_\omega) = 2\nu(P-Q)(a^2/r^2)\cos 2\omega.$$

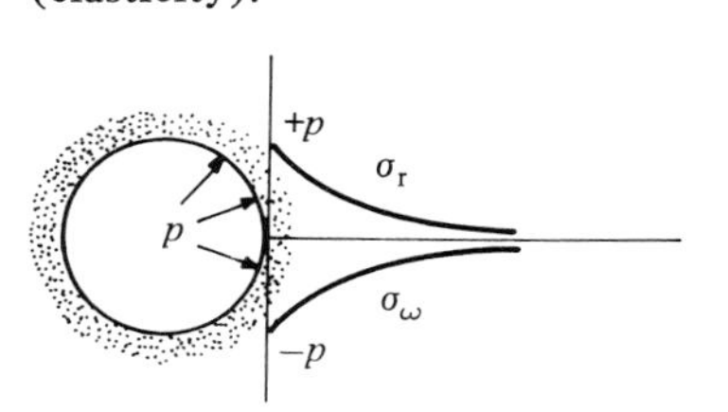

Fig. 12.3. Distribution of σ_r and σ_ω around a circular gallery subjected to an internal pressure p (elasticity).

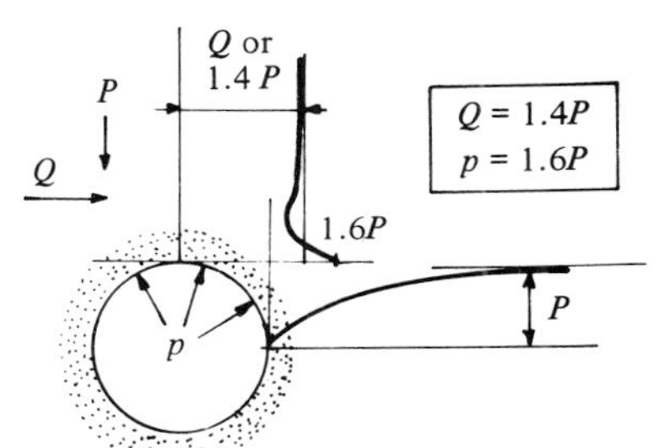

Fig. 12.4. Superimposition of different cases of loading (elasticity).

After the gallery has been opened the stresses on the internal wall are thus:

$$\sigma_r = \tau_{r\omega} = 0,$$
$$\sigma_\omega = P + Q + 2(P - Q)\cos 2\omega,$$
$$\sigma_z = Q + 2\nu(P - Q)\cos 2\omega.$$

These are the most dangerous and it suffices to verify their compatibility with the elastic limit (Mandel, 1959).

In a horizontal gallery, at a depth h, with $P = \gamma h$ and $Q = \kappa\gamma h$ $(0 < \kappa < 1)$, at the side walls $(\omega = 0)$ one has:

$$\sigma_\omega = P(3 - \kappa),$$
$$\sigma_z = P[\kappa(1 - 2\nu) + 2\nu].$$

It can be seen that $\sigma_\omega > \sigma_z > 0$ since $0 < \kappa < 1$ and $0 < \nu < \frac{1}{2}$. It is enough then that $(3 - \kappa)P < R_C$, where R_C is resistance to simple compression taken as an evaluation of the elastic limit. In the same way one has at the keystone $(\omega = \frac{1}{2}\pi)$:

$$\sigma_r = \tau_{r\omega} = 0,$$
$$\sigma_\omega = P(3\kappa - 1),$$
$$\sigma_z = P[(1 + 2\nu)\kappa - 2\nu],$$

and it can be seen that $\sigma_\omega = 0$ for $\kappa = \frac{1}{3}$ and that $\sigma_z = 0$ for $\kappa = 2\nu/(1 + 2\nu)$.

Existing fissures may open at the keystone when κ is lower than the larger of these two values. If κ is larger than these two values the stresses σ_ω and σ_z at the keystone are positive; there are no fissures, or at least those that exist do not open.

The displacements can be calculated from the variation in stresses. In the general case one ends up with very complex formulae. In the simple case of the vertical shaft, and letting $P = Q = \kappa\gamma z$, the variations in the stresses are:

$$\sigma'_r = -Q a^2/r^2,$$
$$\sigma'_\omega = Q a^2/r^2,$$
$$\sigma'_z = \tau'_{r\omega} = 0,$$

and the displacements are:

$$u = Q a^2/2\mu r \quad \text{(radial)},$$
$$v = 0 \quad \text{(tangential)},$$

where μ is the second Lamé coefficient:

$$\mu = E/2(1 - \nu).$$

There is an assumption implicit in the preceding calculation that the initial pressure P is constant, in other words that the variations of stresses brought about by gravity are negligeable in the vicinity of the gallery. The results indicated thus give a good approximation of the reality when one is a long way from the free surface.

12.3 Analysis of the plastic field

Before a gallery completely caves in plastic adaptations appear at various points. These confined plastic deformations are not necessarily dangerous provided that the corresponding radial deformations are not large enough to compromise the use of the tunnel.

All the plastic solutions simplify the problem by assuming the pressure around the gallery to be uniform: thus $P = Q$. These solutions are based upon the assumption of the existence of an annular zone of plasticity around a circular cavity whose equilibrium is ensured by an internal pressure (pressure of the supports, or the pressure $H = c/\tan \varphi$ corresponding to the cohesion). The value of this pressure determines the size of the plastic zone.

12.3.1 *Cylindrical cavity in a weightless elasto-plastic medium*

Let us describe first of all the most complete elasto-plastic mechanism for which the theoretical solution is now known (Salençon, 1966, 1969). The cylindrical circular cavity of actual radius a is situated in a weightless elasto-plastic mass (of characteristics E, ν, c, φ), (fig. 12.5).

The elastic mass is subjected to a pressure P at infinity. Imagine that at the initial instant the cavity is subjected to an internal pressure $p_0 = P$. The field of stresses is then uniform and this situation represents the state of the mass before the excavation of the cavity. At that moment the radius of the cavity is a_0. Let us see now how the whole evolves if p is varied. Two cases should be envisaged, related to active and passive earth pressure.

If p increases from p_0 at first, elastic deformations occur whose expressions are given in section 12.2. From a certain threshold, p_1, the conditions for rupture are reached at the edge of the cavity and plastic deformations appear in a circle of radius b around the cavity (fig. 12.6); a, b and b/a are increasing functions of p but b/a stretches towards a certain limit beyond which the displacements become infinite ($a \to \infty, b \to \infty$). The corresponding limit pressure p_1 represents the largest pressure that one could hope to contain in the cavity.

If p decreases from p_0 elastic deformations appear first of all, obviously associated with a contraction of the cavity. Then a threshold p_2 is reached beyond which rupture occurs by tangential compression at the walls of the cavity. Then a circle of plasticity of radius b develops around the cavity and b increases as p continues to decrease (fig. 12.7). If P is sufficiently large (or if c is sufficiently small) p_2 is positive. In the case where p_2 is negative it signifies simply that there is a depression in the interior of the cavity. Finally, the cavity is completely crushed when

$$p = -H = -c/\tan\varphi.$$

In the same way it can be seen that a cavity excavated in a purely cohesive medium ($\varphi = 0$) can never be completely closed again, whatever the value of P ($p = -\infty$).

The following table summarises the different possibilities:

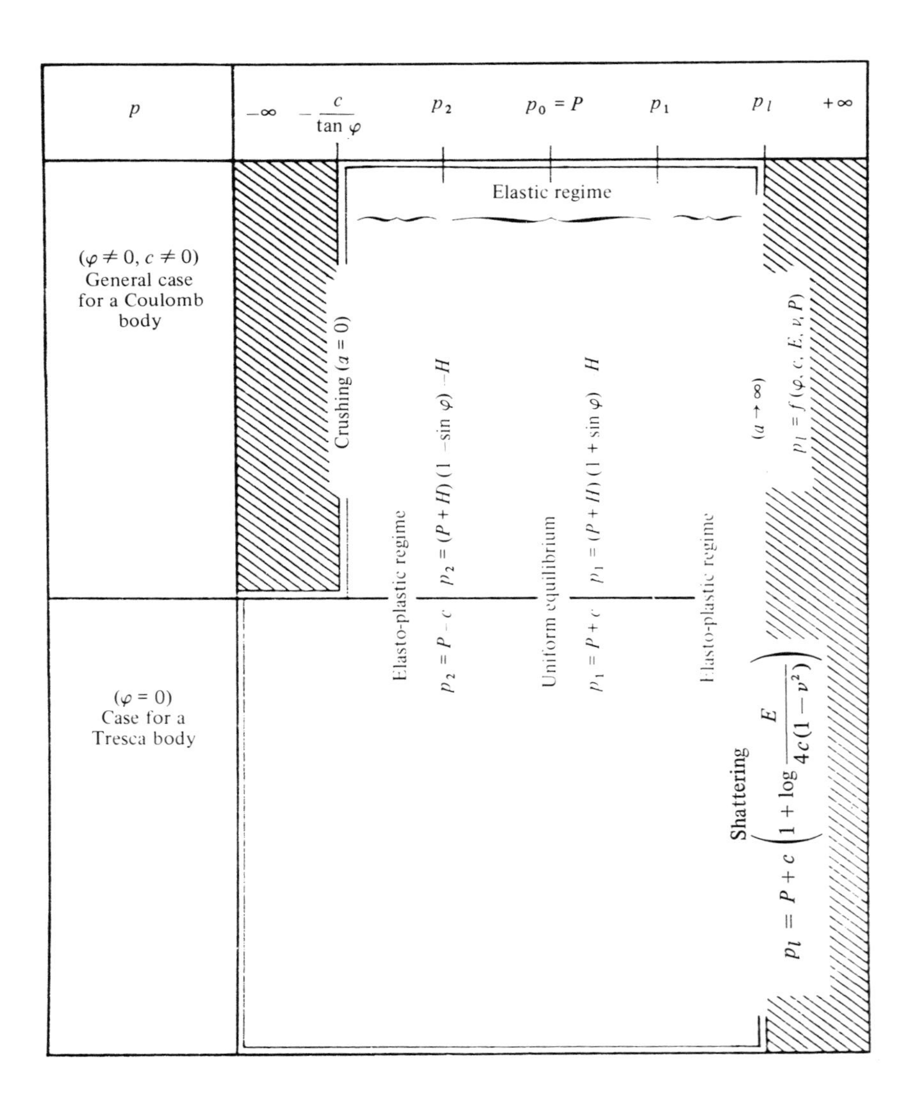

Fig. 12.8 indicates the shape of the curve $a\ (p)$ in the case of Coulomb's body.

We shall not enter into an exposition of the calculations, which are very long, but will restrict the discussion to show, simply, how the problem is stated.

In the plastic zone ($a < r < b$) the elemental equilibrium conditions are written

$$\frac{d\sigma_r}{dr} + \frac{\sigma_r - \sigma_\omega}{r} = 0, \tag{12.1}$$

and σ_ω is eliminated between this equation and that of the conditions for plasticity. Using Coulomb's criterion we get

$$\sigma_\omega + H + j(\sigma_r + H) \tag{12.2}$$

Fig. 12.5. Circular gallery in plasticity: notations.

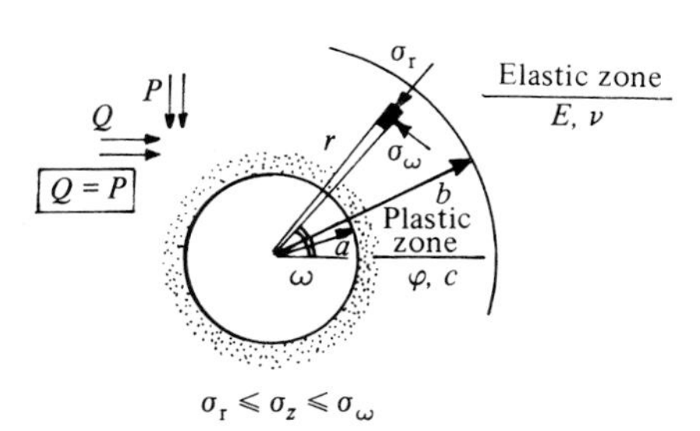

Fig. 12.6. Yielding around an expanding circular gallery.

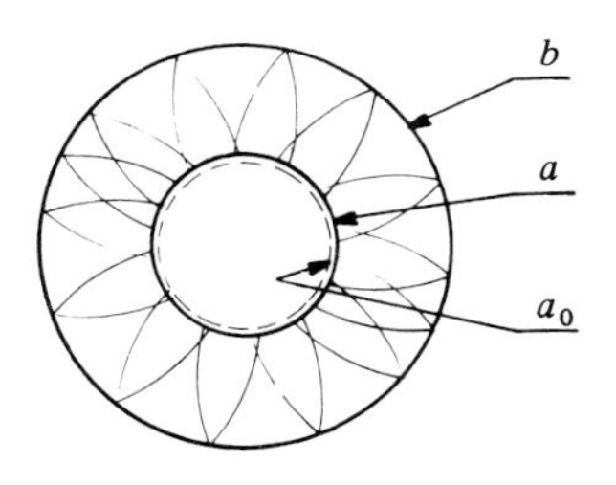

Fig. 12.7. Yielding around a contracting circular gallery.

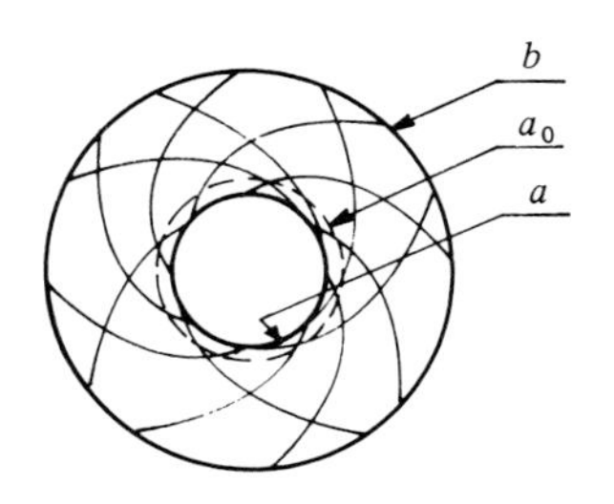

Fig. 12.8. Variation in the radius of a gallery as a function of the internal pressure (elasto-plasticity – weightless mass).

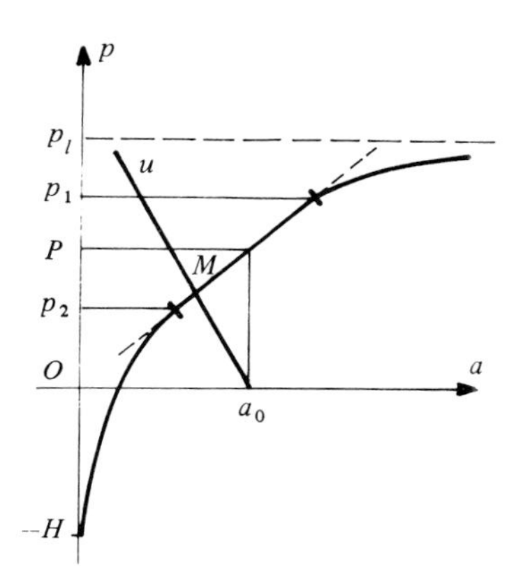

For the case of crushing $j = \tan^2(\frac{1}{4}\pi + \frac{1}{2}\varphi)$. After elimination, integration gives

$$\sigma_r + H = Cr^{j-1}.$$

The constant of integration C is determined by the boundary condition $\sigma_r = p$ for $r = a$. Hence

$$\sigma_r + H = (p+H)(r/a)^{j-1}.$$

It now remains to determine the size of the plastic zone. Letting $A = P + H$ one can write for the elastic domain:

$$\sigma_r + H = A - B/r^2,$$
$$\sigma_\omega + H = A + B/r^2.$$

In order to calculate the radius of the plastic zone and the unknown constant B one must write the continuity between the plastic zone and the elastic zone; from this one finds the relation which determines b:

$$\left(\frac{a}{b}\right)^{j-1} = \frac{P+H}{p+H}(1-\sin\varphi). \tag{12.3}$$

In particular, if $b = a$ (no plastic zone) relation (12.3) gives p_2. Fig. 12.9 gives a graphical representation of σ_r and σ_ω when the plastic zone exists. Equation (12.3) shows that if the medium is cohesive and $p = 0$ one can always imagine a stable circular gallery provided that b/a is large enough.

This result is completely different from that of the theory of elasticity where the stresses only depend upon the shape of the cavity and not on its size. Equation (12.3) allows one to calculate the pressure p within a cavity to maintain a plastic zone of radius b or to avoid the formation of a plastic zone.

When the cavity is excavated in a purely cohesive medium ($\varphi = 0$) one eliminates σ_ω between the relation (12.1) and Tresca's criterion:

$$\sigma_\omega - \sigma_r = 2c \quad \text{(case of crushing);} \tag{12.2'}$$

Fig. 12.9. Distribution of σ_r and σ_ω around a circular gallery showing a plastic ring (contraction).

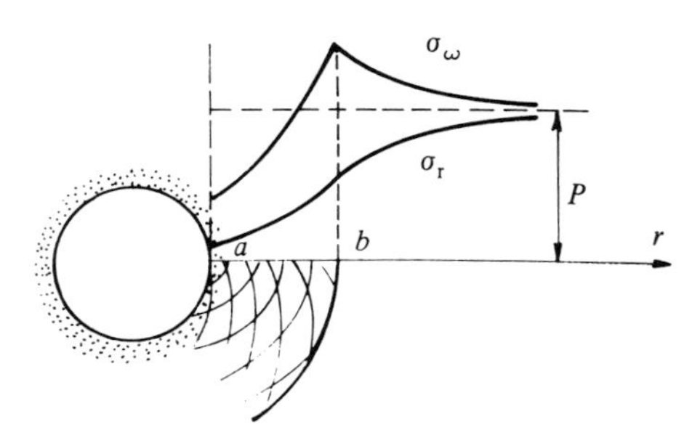

and to calculate the limit pressure p_1 in the case of expansion

$$\sigma_r - \sigma_\omega = 2c \quad \text{(case of expansion)}. \tag{12.2''}$$

Integration gives in the second case

$$\sigma_r = p - 2c \log (r/a).$$

Then

$$\sigma_\omega = p - 2c\,[1 + \log (r/a)].$$

For $r = b$ it is necessary that $\sigma_r + \sigma_\omega = 2P$. Hence

$$P - p = -c\,[1 + 2 \log (b/a)]. \tag{12.4}$$

If one puts $b = a$ into equation (12.4) the appearance of the plastic zone gives $p_1 = P + c$.

In equation (12.4) it was proved that the limit of b/a is: $E/4c(1 - \nu^2)$ as p increases. By incorporating this value, the limit pressure is obtained as

$$p_1 = P + c\left(1 + 2 \log \frac{E}{4c(1-\nu^2)}\right).$$

It will be noticed that the formulae for the transition into the plastic domain do not use the soil characteristics E and ν, although these do appear in the expression for limit pressure in expansion.

In order to calculate $a = f(p)$ the constitutive equation of the material, especially that for dilatancy must be taken into account. The calculation is very long (Salençon, 1966 and 1969). It should be remembered that this results in explicit formulae of the form $a = f(p, P, a_0, \varphi, c, E, \nu$, dilatancy) which are difficult to present on account of the large number of parameters but which are easily handled on a computer.

12.3.2 *Calculating the lining of underground structures*

The preceding formulae seem to be perfectly applicable to vertical shafts. Compared to a vertical cavity, a horizontal layer between the depths of h and $h + \mathrm{d}h$ represents a weightless mass. Unfortunately, the value of P, the stress extending to infinity, or the initial stress, or again the earth pressure at rest, is not well known. It is generally accepted that it lies between $\frac{1}{2}\gamma h$ and γh, this latter value being acceptable for very thick clays. But if orogenic movements take place P can be different from γh and even the hypothesis $P = Q$ (orthotropy of the field of stresses extending to infinity) becomes suspect. It is almost always incorrect in a rock mass where the regional tectonics impose a state of stress that is more often than not anisotropic.

All the preceding formulae assume σ_z to be the intermediate principal stress.

This hypothesis is almost always true but it should be verified every time, whether for shafts or for other underground structures which we shall now discuss.

Since the underground structure is most often a horizontal gallery, the stresses to infinity (that is the stresses at a few diameters distance from the excavation, or again the initial stress in the soil) are different: $P = \gamma h$ for the vertical stress and $Q = \kappa P$ for the horizontal stress. In soils $\kappa = 1$ is always a pessimistic hypothesis, which means that the preceding formulae can be used while remaining on the safe side. The validity of these formulae will be even better when the gravity gradient exercises little influence, that is when the ratio a_0/h is small.

The efficiency of a lining can be estimated by envisaging a tube of radius a_0 placed in the mass before the underground structure is opened up. An external pressure forces the tube to contract, and as a rule, one hopes that it stays in the elastic state. The straight dashed line a_0u in fig. 12.8 represents the elastic deformation curve $a(p)$ of a tube; the intersection of M with the curve $p(a)$ gives both the load acting on the lining and the radial deformation.

One might very well deduce that it is beneficial to use very deformable linings since they support the smallest of loads. This reasoning is not however totally satisfactory – firstly because very deformable bodies are often the least resistant (even if this comment does not have overall bearing it does apply at least to particular types, e.g. concrete) and secondly large contraction deformations bring about a loosening of the soil, which means a lowering of the angle of internal friction in the plastic ring and consequently an increase in the active earth pressure. This danger is particularly likely when the stress/strain curve of the soil has a well-defined peak. Finally, because in practice the reality on site does not correspond entirely to the model in Fig. 12.8; the lining is not installed immediately after the gallery has been bored, and is not fastened rigorously against the earth. The position of the straight line a_0u in fig. 12.8 is thus completely theoretical. Thus, if the space between the soil and the lining is too large it is advisable to fill it in, by means of cement grouting for instance, which, if carried out under pressure, removes the harmful effects of decompression. Further, grouting, while making the contact with the soil uniform, also avoids the risk of loading the tube asymmetrically and thus causing the lining to take on an oval shape and buckle. Generally one rarely exceeds pressures of 0.5 to 0.6 MPa in present-day works.

12.3.3 *Collapse of an underground structure*

Caquot and Kerisel (1966) proposed a collapse formula for an underground structure in a heavy Coulomb material, based on the idea that rupture occurs when the plastic zone reaches the free surface, in other words when the plastic flow ceases to be confined and becomes free. In this case the movement may persist until the work is completely destroyed. With these hypotheses one has $b = h$ with $P = \gamma h$.

The calculation is done with a certain number of simplifications: it neglects the elasticity of the soil; it is nearly valid and on the safe side above the gallery (that is to say in the most dangerous zone); one writes that the limit equilibrium is reached at every point situated on the vertical of the keystone of the arch. This calculation permits one to define the radial stress p that the lining should be able to support. Using the same notation as before one arrives at the formula:

$$p = \frac{\gamma h}{j-2}\left[\frac{a_0}{h} - \left(\frac{a_0}{h}\right)^{j-1}\right] - H\left[1 - \left(\frac{a_0}{h}\right)^{j-1}\right].$$

When the cohesion is zero, the second term may be omitted. When $\varphi = 0$ there is an element of uncertainty which can be removed, giving

$$p = \gamma a_0\left(\frac{h}{a_0} - 1\right) - 2c \log\frac{h}{a_0}.$$

Despite the visible imperfections of the hypotheses and particularly the impossibility of calculating the radial deformation, Caquot's formulae are interesting for they result in simple explicit values, neglecting the soil parameters E and ν whose role is of relatively little importance when $b = h$ is large.

A more rigorous solution for the collapse of an underground structure in a rigid, heavy, plastic mass has been given by d'Escatha and Mandel (1971) using the method of characteristics and numerical calculation on the computer.

12.4 Underground structures subjected to infiltrations

An underground structure often serves, involuntarily, as a drain for the mass it traverses; the water flows towards the gallery and the volume forces produced by the seepage forces are always a factor of instability. Such an example is the unstable gallery in a mine whose stability is improved when the gallery is left to be deluged. The infiltrations, of course, originate most often from privileged passages or from particular layers so that the following calculations worked out in the framework of axial symmetry, with radial flows and for circular galleries are extremely idealised and their interpretation should be examined both geologically and geotechnically at the time of execution.

12.4.1 *Radial convergent flow*

If an impermeable lining is placed around the gallery, it supports not only the stresses due to the weight of the earth, but also the hydrostatic load of the water table. If the soil is treated prior to the boring so that it is more watertight in a circular aureole of radius R concentric to the gallery, the situation is more favourable since the loss of head of water occurs at a certain distance from the lining and a part of the stress so produced is supported by the soil (with characteristics φ and c) between a, the radius of the gallery, and R (fig. 12.10).

Let us study the radial convergent flow of the water from the soil (considered as a source) to the excavation (considered as a well) that is to say it is assumed that all the loss of head is localised between a and R. The hydraulic gradient $i = \mathrm{d}h/\mathrm{d}r$ varies along a radius, but the discharge $\mathrm{d}q$ in a tube is obviously constant; one has thus:

$$\mathrm{d}q = V\mathrm{d}S = k\frac{\mathrm{d}h}{\mathrm{d}r} r\,\mathrm{d}\omega\,\mathrm{d}z = C,$$

or again

$$\mathrm{d}h = \lambda\frac{\mathrm{d}r}{r}.$$

Hence

$$h = \lambda \log(r/C_1),$$

where C_1 is a constant of integration. For $r = a$ one has $h = 0$ thus $C_1 = a$ and $h = \lambda \log(r/a)$. For $r = R$ one has $h = \mathscr{H}$ (on average) thus

$$\lambda = \frac{\mathscr{H}}{\log(R/a)}$$

Fig. 12.10. Circular gallery subjected to infiltrations: notations.

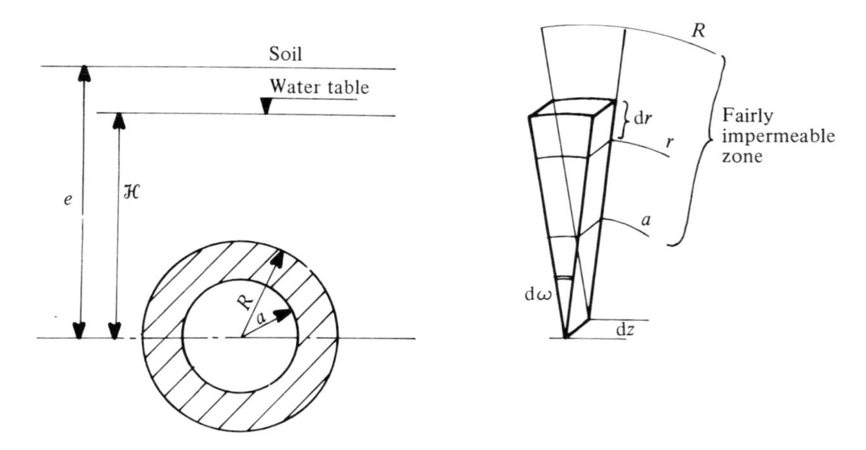

and

$$h = \frac{\mathscr{H}}{\log (R/a)} \log (r/a).$$

Hence finally

$$\frac{\mathrm{d}h}{\mathrm{d}r} = \mathrm{i} = \frac{\mathscr{H}}{r \log (R/a)}.$$

As the rupture of an underground structure always takes place at the keystone this approximation can be improved by taking into account the hydraulic pressure at the keystone, which is $\mathscr{H} - a$ for $r = R$. The gradient then becomes

$$\mathrm{i} = \frac{\mathscr{H} - a}{r \log (R/a)} .$$

This new expression shows moreover the flow of the infiltrations regularises in some way the instability since the hydraulic gradient is larger on the floor and smaller at the keystone.

12.4.2 *Equilibrium of an underground structure affected by infiltrations*

It will be assumed that the structure is stable as long as the circle (a, R) has not been broken, and the calculation will take place, as for Caquot's formula, in the most dangerous zone, that is to say around the vertical ascendent radius. The elementary volume ΔV (r dr dω dz) is subjected to radial stresses σ_r and tangential stresses σ_ω, to its own weight and to the seepage force. By projecting on the vertical axis we find

$$\sigma_\omega \,\mathrm{d}r\,\mathrm{d}z\,\mathrm{d}\omega - \mathrm{d}(\sigma_\mathrm{r} r \,\mathrm{d}\omega\,\mathrm{d}z) - \gamma_\mathrm{i} r\,\mathrm{d}\omega\,\mathrm{d}r\,\mathrm{d}z - \gamma_\mathrm{w}\,\mathrm{i} r\,\mathrm{d}\omega\,\mathrm{d}z\,\mathrm{d}r = 0$$

(γ_i is the submerged density and $\mathrm{i}\gamma_\mathrm{w} \Delta V$ is the seepage force as indicated in chapter 10) or again

$$\sigma_\omega - \sigma_\mathrm{r} - \gamma_\mathrm{i} r - \gamma_\mathrm{w}\,\mathrm{i} r - r\frac{\mathrm{d}\sigma_\mathrm{r}}{\mathrm{d}r} = 0.$$

In making the hydraulic gradient explicit, the equilibrium equation becomes

$$\sigma_\omega - \sigma_\mathrm{r} - \gamma_\mathrm{i} r - \gamma_\mathrm{w}\frac{\mathscr{H} - a}{\log (R/a)} - r\frac{\mathrm{d}\sigma_\mathrm{r}}{\mathrm{d}r} = 0. \tag{12.5}$$

To determine the pressure on the lining it suffices to write that the conditions for rupture,

$$\sigma_\omega + H = j(\sigma_\mathrm{r} + H) \tag{12.6}$$

are verified on the vertical radius between a and R. Eliminating σ_ω between (12.5) and (12.6) gives

$$(j-1)(\sigma_r + H) - \gamma_i r - \gamma_w \frac{\mathscr{H} - a}{\log (R/a)} - r\frac{d\sigma_r}{dr} = 0, \tag{12.7}$$

and letting

$$\sigma_r + H = y \quad \text{and} \quad A = \gamma_w \frac{\mathscr{H} - a}{\log (R/a)}, \tag{12.8}$$

we find

$$ry' - (j-1)y + \gamma_i r + A = 0. \tag{12.7$'$}$$

The solution of

$$ry' - (j-1)y = 0$$

is $y = Cr^{j-1}$. Hence

$$y' = (j-1)Cr^{j-2} + C'r^{j-1},$$

and (12.7′) becomes

$$\frac{dC}{dr} r^j + \gamma_i r + A = 0.$$

Hence

$$C = -\frac{\gamma_i}{2-j} r^{2-j} - \frac{A}{1-j} r^{1-j} + B,$$

where B is a constant of integration. Finally, the solution of (12.7) is:

$$\sigma_r + H = Br^{j-1} + \frac{\gamma_i}{j-2} r + \frac{A}{j-1}.$$

To determine the value of the constant of integration B, let us write that above the zone considered, at the keystone, the stress is equal to the weight of the earth. To simplify the equation, not that this is strictly necessary, let us assume that $e = \mathscr{H}$ (which means that the water table is at the soil surface).

For $r = R$ one thus has $\sigma_r = \gamma_i(e - R)$. Hence

$$B = \left(\gamma_i(e-R) + H - \frac{\gamma_i R}{j-2} + \frac{A}{j-1}\right)\left(\frac{1}{R^{j-1}}\right)$$

and

$$\sigma_r = -H + \left(\gamma_i(e-R) + H - \frac{\gamma_i R}{j-2} - \frac{A}{j-1}\right)\left(\frac{r}{R}\right)^{j-1} + \frac{\gamma_i}{j-2} r + \frac{A}{j-1}.$$

Finally the pressure on the support is

$$p = \sigma_{r(a=r)} = -H + \left(\gamma_i(e-R) + H - \frac{\gamma_i R}{j-2} - \frac{A}{j-1}\right)\left(\frac{a}{R}\right)^{j-1}$$

$$+ \frac{\gamma_i}{j-2} a + \frac{A}{j-1} . \qquad (12.9)$$

A numerical application easily shows that the preponderant term in $p + H$ is almost always the last term of (12.9) when the underground structure is deep.[1]

1 Take for example the following data $a = 1$ m; $R = 3$ m; $e = \mathscr{H} = 20$ m; $\varphi = 30°$; $j = 3$; $c = 0.58 \times 10^5$ Pa; $H = 10^5$ Pa; $\gamma_i = 10^4$ N/m^3; $\gamma_w = 10^4$ N/m^3. Equation (12.8) gives $A \approx 1.8 \times 10^5$ Pa and the different terms of equation (12.9) have the following numerical values:

$$p = [-1 + (1.7 + 1 - 0.3 + 0.9)\tfrac{1}{9} + 0.1 + 0.9]\ 10^5 \text{ Pa} = 0.37 \times 10^5 \text{ Pa}.$$

REFERENCES AND FURTHER READING

Chapter 1

A.S.T.M. (1946) Tentative definition of terms and symbols relating to soil mechanics. *Book of American Society for Testing Materials and Materials Standards,* D653-42T, part II, p. 1473.

Bureau of Reclamation (1963) *Earth Manual.* US Department of the Interior, Denver, Colorado.

Casagrande, A. (1947) Classification and Identification of Soils. *Proc. ASCE,* June, pp. 783–810.

Dearman, W.R. (1974) The characterization of rock for civil engineering practice in Britain. In *La Géologie de L'Ingénieur* (ed. L. Calembert), Société Géologique de Belgique, Liège, Belgium, pp. 1–75.

Glossop, R. and Nixon, I.K. (1961) Site investigations. In *Civil Engineering Reference Book* (ed. J. Comrie), 2nd edn, Butterworths, London.

Golder, H.Q. and Meigh, A.C. (1961) Soil mechanics. In *Civil Engineering Reference Book* (ed. J. Comrie), 2nd edn, Butterworths, London.

Habib, P. (1958) La dispersion des résultats des essais de sol, *Annales ITBTP,* October.

International Society for Soil Mechanics (1981) *Lexicon in 8 languages. Technical terms, symbols and definitions.* The Bryant Press Limited. Toronto.

Keefe, H.G. and Elsden, O. (1964) Site investigations. In *Hydro Electric Engineering Practice* (ed. G. Brown), 2nd edn, Butterworths, London pp. 249–82.

Mayer, A. (1947), *Sols et Fondations,* Armand Colin, Paris.

Peltier, R. (1969) *Manuel du Laboratoire Routier,* Dunod, Paris.

Richey, J.E. (1964) *Elements of Engineering Geology,* Pitman and Sons, London, pp. 1–157.

Rowe, P.W. (1972) The relevance of soil fabric to site investigation practice. 12th Rankine Lecture 1972, *Géotechnique,* **22,** 2, 195–300.

Schneebeli, G. (1966) *Hydraulique Souterraine,* Eyrolles.

Taylor, D.W. (1956) *Fundamentals of Soil Mechanics,* Wiley, New York.

Terzaghi, K. (1943) *Theoretical Soil Mechanics,* Wiley, New York, pp. 1–510.

Terzaghi, K. and Peck, R. (1948) *Soil Mechanics in Engineering Practice,* Wiley, New York.

Chapter 2

Bishop, A.W. and Bjerrum, L. (1960) The relevance of the triaxial test to the solution of stability problems. *A.S.C.E. Research Conference on Shear Strength of Cohesive Soils,* Boulder, Colorado, pp. 437–501.

Bishop, A.W. and Henkel, D.J. *The Measurement of Soil Properties in the Triaxial Test,* 2nd edn. Edward Arnold, London, pp. 1–225.

Donath, F.A. (1966) A triaxial pressure apparatus for testing of consolidated or unconsolidated materials subjected to pore pressure. In *Testing Techniques for Rock Mechanics,* American Society for Testing Materials and Materials Standards, publication no. 402.

Habib, P. (1953) La résistance au cisaillement des sols. *Annales ITBTP,* January.

Marsal, R.J. (1965) *Investigacion Sobre el Comportamiento de Suelos Granulares y Muestras de Enrocamiento,* Comision Federal de Electricidad, Mexico (D.F.), February.

Sanglerat, G. (1972) *The Penetrometer and Soil Exploration,* Elsevier.

Skempton, A.W. (1960) Effective stress in soils, concrete and rocks. *Conference on Pore Pressure and Suction in Soils,* Butterworths, London.
Taylor, D.W. (1956) *Fundamentals of Soil Mechanics,* Wiley, New York.
Terzaghi, K. and Peck, R. (1948) *Soil Mechanics in Engineering Practice,* Wiley, New York.

Chapter 3

Caquot, A. and Kérisel, J. (1966) *Traité de Mécanique des Sols,* Gauthier-Villars, Paris.
Lundgren, H. (1957) Dimensional analysis in soil mechanics, *Acta Polytechnica,* Copenhagen, no. 237.
Mandel, J. (1962) Essais sur modèles réduits en mécanique des terrains. Étude des conditions de similitude, Revue de L'Industrie Minerale, September, vol. 44, no. 9.
Oberti, G. (1957) Arch dams: development of model researches in Italy, *Proc. ASCE,* August.
Roscoe, K.H. (1970) The influence of strains in soil mechanics, *Géotechnique,* **20,** 2, 129–70.
Terzaghi, K. (1943) *Theoretical Soil Mechanics,* Wiley, New York.
Zelikson, A. (1967) Représentation de la pesanteur par gradient hydraulique dans les modèles réduits en géotechnique, *Annales ITBTP,* November, pp. 1556–80.

Chapter 4

Bishop, A.W. (1971) The influence of progressive failure in the method of stability analysis, *Geotechique,* **21,** 168–72.
Caquot, A. and Kérisel, J. (1966) *Traité de Mécanique des Sols,* Gauthier-Villars, Paris.
Chowdhury, R.N. (1978) Slope analysis. *Developments in Geotechnical Engineering,* vol. 22, Elsevier Scientific Publishing Company, Amsterdam.
Kennedy, B.A. and Niermeyer, K.E. (1970) Slope monitoring system used in the prediction of major slope failures at the Chugincamata mine, Chile, *Proc. Symposium on Planning and Open Pit Mines,* Johannesburg.
Logeais, L. (1982) *Pathologie des murs de soutenement*, Editions du Moniteur, Paris.
Morgenstern, N.R. (1977) State of the art report and preliminary general report on slopes and excavations, *9th Conf. Soil Mech. Found. Eng.* Tokyo, vol. 2, pp. 547–603.
Post, G. and Londe, P. (1953) *Les Barrages en Terre Compactée,* Gauthier-Villars, Paris.
Radenkovic, D. (1962) *Théorie des Charges Limites. Extension à la Mécanique des Sols* (Séminaire de Plasticité, École Polytechnique), Scientific and technical publications of the Air Ministry, Paris.
Skempton, A.W. (1977) Slope stability of cuttings in brown London clay, *9th Conf. Soil Mech. Found. Eng.* Tokyo, vol. 3, pp. 261–70.
Taylor, D.W. (1956) *Fundamentals of Soil Mechanics,* Wiley, New York.
Vidal, H. (1966) La terre armée, *Annales ITBTP,* nos. 223–4.

Chapter 5

Coquand, R. (1972) *Routes,* Eyrolles, Paris.
Jeuffroy, G.A. (1967) *Conception et Construction des Chaussées,* Eyrolles, Paris.
L.C.P.C. (1965) Essais de plaques et mécanique des chaussées, *Bulletin de Liaison des Laboratoires Routiers* (L.C.P.C. supplement, February).
R.R.L. (1952) *Soil Mechanics for Road Engineers,* Road Research Laboratory (Dep. of Scientific and Industrial Research), London.

Woods, K.B. (1960) *Highway Engineering Handbook,* McGraw-Hill, New York.
Zemmour, P. (1966) Granularité des assises stabilisées mécaniquement, Bulletin de Liaison des Laboratoires Routiers (L.C.P.C. no. 18, March/April).

Chapter 6

Lewis, W.A. (1954) *Further Studies in the Compaction of Soil and the Performance of Compaction Plant,* Road Research Technical Paper, no. 33, London.
Peltier, R. (1969) *Manuel du Laboratoire Routier,* Dunod, Paris.
Post, G. and Londe, P. (1953) *Les Barrages en Terre Compactée,* Gauthier-Villars, Paris.
Proctor, R.R. (1933) The design and construction of rolled earth dams, *Engineering New Record,* August/September.

Chapter 7

Broms, B. and Bennermark, H. (1967) Stability of clays at vertical openings, *Proc. ASCE,* January.
Giraudet, P. (1965) Recherches expérimentales sur les fondations soumises à des efforts inclinés ou excentrés, *Annales Ponts et Chaussées,* May/June, p. 167.
Giroud, J.P., Tran-Vo-Nhiem and Obin, J.P. (1973) *Tables pour le Calcul des Fondations,* Dunod, Paris, vol. 3.
Habib, P. (1961) Force portante et déformations des fondations superficielles, *Annales ITBTP,* July/August.
Habib, P. and Suklje, L. (1954) Étude de la stabilité des fondations sur une couche d'épaisseur limitée, *Annales ITBTP,* November.
Logeais, L. (1971) Pathologie de la construction: les fondations superficielles, *Bâtir,* February.
Lundgren, H. and Mortensen, K. (1953) Determination by the theory of plasticity of the bearing capacity of continuous footings in sand, *Proc. 3rd Int. Conf. Soil Mech.,* Zurich, vol. 1, p. 409.
Mandel, J. (1942) Équilibre par tranches planes des solides à la limite d'écoulement, Ph.D. Thesis, University of Lyon.
Mandel, J. (1965) Interférence plastique des semelles filantes, *Proc. 6th Int. Conf. Soil Mech.,* Montreal, vol. 1, p. 127.
Mandel, J. and Salençon, J. (1969) Force portante d'une fondation sur un bi-couche, *Proc. 7th Int. Conf. Soil Mech.,* Mexico.
Meyerhof (1956) Discussion of 'rupture surfaces in sand under oblique loads', *Proc. ASCE,* Journal of the Soil Mechanics Division, July.
Salençon, J. (1965) Force portante d'une fondation peu profonde. Emploi de la méthode des caractéristiques, *Annales Ponts et Chausséés,* May/June, p. 195.
Skempton, A.W. (1951) The bearing capacity of clay. *Proc. Building Research Congress,* div. I, part III, p. 180.
Sokolovski, V. (1960) *Statics of Soil Media,* London, 1960.
Terzaghi, K. and Peck, R. (1957) *Mécanique des Sols Appliquée*, Dunod, Paris.

Chapter 8

Boussinesq, J. (1885) *Application des Potentiels à l'Etude de l'Equilibre et du Mouvement des Sols Elastiques,* Gauthier-Villars, Paris.
Burmister, D. (1956) Stress and displacement characteristics of a two-layer rigid base soil system: influence diagrams and practical application, *Proc. Highway Research Board.*
Giroud, J.P. (1972) *Tables pour le Calcul des Fondations,* Dunod, Paris, vols. 1 and 2.

Habib, P. and Luong, M.P. (1972) Force portante des constructions de grande hauteur, *Annales ITBTP,* March, pp. 51–9.
Habib, P. and Puyo, A. (1970) Stabilité des fondations des constructions de grande hauteur, *Annales ITBTP,* November, pp. 118–24.
Leonards, G.A. (1968) *Les Fondations* (by a group of American authors), Dunod, Paris.
Mandel, J. (1957) Consolidation des couches argileuses *Proc. 4th Int. Conf. Soil Mech.,* London, vol. 1, p. 360.
Mandel, J. (1961) Tassements produits par la consolidation d'une couche d'argile de grande épaisseur, *Proc. 5th Int. Conf. Soil Mech.,* Paris, vol. 1, p. 733.
Mathian, J. (1972) Méthode d'observation des mouvements verticaux et de représentation des résultats. Le comportement des sols avant la rupture. Journées du C.F.M.S., 17–19 May, 1971, *Bulletin de Liaison des L.P.C.,* June.
Newmark, N.W. (1942) Influence charts for computation of stresses in elastic foundations. *Eng. Exp. Station Bulletin,* University of Illinois, November.
Skempton, A.W. and Bjerrum, L. (1957) A contribution to the settlement analysis of foundations on clay. *Géotechnique,* 7, 168.
Skempton, A.W. and MacDonald, J.R. (1956) *The Allowable Settlement of Buildings,* Institution of Civil Engineers, London, December.
Taylor, D.W. (1948) *Fundamentals of Soil Mechanics,* Wiley, New York.
Timoshenko, S. and Goodier, J. (1961) *Théorie de l'Elasticité*, Béranger, Paris.

Chapter 9

Bureau Sécuritas (1977) *Recommendations Concernant les Tirants d'Ancrage,* Eyrolles.
Cambefort, H. (1964) Essai sur le comportement en terrain homogène des pieux isolés et des groupes de pieux, *Annales ITBTP,* December.
Caquot, A. and Kérisel, T. (1966) *Traité de Mécanique des Sols,* Gauthier-Villars, Paris.
Document technique unifié 13.2 (1978). Travaux de fondations profondes pour le bâtiment. (Cahier des charges, Prescriptions communes, Clauses spéciales, Mémento), *CSTB,* Paris, June.
Habib, P. (1953) Essais de charge portante de pieux en modèle réduit; dans 'La mécanique des sols et la force portante des pieux', *Annales ITBTP,* March/April.
Habib, P. (1960) Le frottement négatif, *Annales ITBTP,* January.
Habib, P. and Puyo, A. (1972) Utilisation en fondation et en génie civil d'un ressort plat pouvant supporter des forces très grandes, *Annales ITBTP,* September.
International Congress Soil Mech. (7th) (1969) Special session no. 14, *Les Parois Moulées dans le Sol*, Mexico.
International Congress Soil Mech. (7th) (1969) Special session no. 15, *Les Ancrages dans le Sol,* Mexico.
Logeais, L. (1981) *La Pathologie des Fondations,* Editions du Moniteur, Paris.
Salençon, J. (1966) Expansion quasi statique d'une cavité à symétrie sphérique ou cylindrique dans un milieu élastoplastique, *Annales des Ponts et Chausées,* no. III; p. 175.
Sanglerat, G. Olivari, G. and Cambou, B. (1983) *Practical Problems of Soil Mechanics,* 2nd edn., Elsevier, Amsterdam.
Tcheng, Y. (1966) Fondations profondes en milieu pulvérulent à diverses compacités, *Annales ITBTP,* March/April.

Chapter 10

Barbedette, R. and Sabarly, F. (1953) Études et utilisation des coulis d'injection argile-ciment, *Proc. 3rd Int. Conf. Soil Mech.,* Zurich, vol. 1, p. 85.

Carlier, M. (1965) *Cours d'Hydraulique Générale: Part 5, Hydraulique Souterraine,* ENGREF, Riber, Paris, pp. 81–95.

Dupeuble, P. and Habib, P. (1969) *Coupure Etanche en Béton Plastique*, 7th Int. Conf. Soil Mech., Mexico, Special session no. 14, SDTBTP, Paris, Communication 14/12.

Elsden, O., Keefe, H.G. and Bishop, A.W. (1964) Embankment dams. In *Hydro Electric Engineering Practice* (ed. G. Brown) 2nd edn. Butterworths, London, vol. 1, pp. 347–479.

Fellenius, W. (1927) *Erdstatische Berechnungen mit Reibung und Kohesion,* Ernst, Berlin.

Fellenius, W. (1936) Calculation of stability of earth dams, *Trans. 2nd Congress Large Dams,* vol. 4, p. 445.

Habib, P. (1954) Application de la méthode des analogies électriques aux problèmes d'hydraulique laminaire, Journées de l'Hydraulique, Alger (published in *La Houille Blanche*).

Habib, P. and Luong, M.P. (1966) Étude théorique et expérimentale de la stabilité des tuyaux et buses cylindriques placés dans les remblais, *Annales ITBTP,* February.

Jaeger, C. (1956) *Engineering Fluid Mechanics,* Blackie and Son, Glasgow, pp. 1–529.

Jürgenson, L. (1934) The application of theories of elasticity and plasticity to foundation problems, *J. Boston Soc. Civ. Eng.*, July.

Louis, C. (1973) *Rock Hydraulics,* Bureau de Recherches Géologiques et Minières, Orleans.

Muskat, M. (1937) *The Flow of Homogeneous Fluids through Porous Media,* New York, London.

Post, G. and Londe, P. (1953) *Les Barrages en Terre Compactée,* Gauthier-Villars, Paris.

Schneebeli, G. (1966) *Hydraulique Souterraine,* Eyrolles, Paris.

Terzaghi, K. and Peck, R. (1957) *Mécanique des Sols Appliquée,* Dunod, Paris, pp. 51–3.

U.S. Department of Interior (1976/77) *Failure of Teton Dam,* Two volume report of Independent Panel (1976) and International Review Group (1977). US Government Printing Office, Washington DC.

Chapter 11

Bernaix, J. (1967/75) *Étude Géotechnique de la Roche de Malpasset,* Paris 1967 and *Industrie Minérale,* 15 December 1975, pp. 45–69.

Bieniawski, Z.T. (1974) Estimating the strength of rock materials, *J. South African Inst. Min. Met.,* **74**, 312–20.

Deere, D.U. (1963) Technical description of rock cores for engineering purposes, *Rock Mech. Eng. Geol.*, **1**, 18–22.

Farran, J. and Thenoz, B. (1965) L'altérabilité des roches, ses facteurs, sa prévision, *Annales ITBTP,* November.

Goguel, J. (1959) *Application de la Géologie aux Travaux de l'Ingénieur,* Masson.

Goodman, R.E. (1980) *Introduction to Rock Mechanics,* Wiley, New York.

Griffith, A.A. (1924) *Theory of Rupture,* 1st Conf. Appl. Mech., Delft, pp. 55–63.

Habib, P. (1950) Détermination du module d'élasticité des roches en place, *Annales ITBTP,* September.

Habib, P. (1972) Comportement comparé des sols et des roches, *Proc. 6th Int. Cong. Rheol.,* Lyon, September.

Habib, P. (1972) Choix de la localisation des mesures de contraintes et techniques d'essais in situ, *Proc. Int. Symp. Construction des Cavités Souterraines,* Lucerne September, Part III.

Habib, P. and Marchand, R. (1952) Mesure des pressions de terrain par l'essai de vérin plat, *Annales ITBTP,* October.

Habib, P. and Bernaix, J. (1966) La fissuration des roches, *Proc. 1st Int. Cong. Rock Mech.,* Lisbon.

Heim, A. (1878) *Mechanismus der Gebingsbildung,* Basle.
Heim, A. (1905) Geologische Nachlose, Tunnelban und Gebirgsdruck, *Vierteljahrschrift der Naturforsch. Gesellschaft,* Zurich, p. 50 and *Schweizerische Banzeitung,* (February 1912) p. 50.
Hoek, E. (1976) State of the art paper on rock slopes. *Proc. ASCE Geotech. Eng. Div. Spec. Conf.,* Boulder, Colorado, vol. 2.
Hoek, E. and Bray, J.W. (1977) *Rock Slope Engineering,* 2nd edn, Inst. Min. Met., London, pp. 1–402.
Jaeger, C. (1979) *Rock Mechanics and Engineering,* 2nd edn, Cambridge University Press, Cambridge, pp. 1–523.
Jaeger, J.C. and Cook, N.G.W. (1969) *Fundamentals of Rock Mechanics,* Methuen.
John, K.W. (1969) Civil engineering approach to evaluate strength and deformation of regularly jointed rock, *Proc. 11th Symp. Rock Mech.,* Berkeley, pp. 69–80.
Jouanna, P. and Rayneau, C. (1971) *Influence de la Température sur l'Écoulement en Milieux Fissurés,* Symp. Soc. Int. Mec. Roches, Nancy, October, section II.4.
Ladanyi, B. and Archambault, G. (1970) Simulation of shear behaviour of a jointed rock mass, *Proc. 11th Symp. Rock Mech.,* AIME, New York, pp. 105–25.
Londe, P. (1965) Une méthode d'analyse à trois dimensions de la stabilité d'une rive rocheuse, *Annales des Ponts et Chausées,* January.
Lugeon, M. (1933) *Barrages et Géologie,* Dunod, Paris.
Mandel, J. (1959) Les calculs en matière de pression de terrains, *Revue de l'Industrie Minérale,* January and April.
Maury, V. (1970) *Mécanique des Milieux Stratifiés,* Dunod, Paris.
Mayer, A. (1965) La mécanique des roches, *Proc. 6th Int. Conf. Soil Mech.,* Montreal, vol. 3, pp. 104–13.
Mayer, A., Habib, P. and Marchand, R. (1951) Mesure en place des pressions de terrains, *Int. Conf. Rock Pressure,* comm. B.6, INICHAR, Liege.
Morlier, P. (1966) Le fluage des roches, *Annales ITBTP,* January.
Müller, L. (1963/77) *Der Felsbau,* Two volumes, Enke Verlag, Stuttgart.
Obert, L. and Duvall, W.I. (1967) *Rock Mechanics and the Design of Structures in Rock,* Wiley, New York.
Stagg, K.G. and Zienkiewics, O.C. (1969) *Rock Mechanics in Engineering Practice,* Wiley, New York.
Talobre, J. (1968) *La Mécanique des Roches,* Dunod, Paris.
Tincelin, E., Léonet, O. and Sinou, P. (1971) Étude du comportement d'un toit en fonction de différents modes de boulonnage, *Proc. Symp. Int. Soc. Rock Mech.,* Nancy, October, section III.6.

Chapter 12

Bernède, J., Habib, P., Plouviez, P. and Stragiotti, L. (1968) Mesure de contraintes dans le revêtement d'un tunnel alpin, *Proc. Symp. Int. Soc. Mech. Rocks,* Madrid.
Caquot, A. and Kérisel, J. (1966) *Traité de Mécanique des Sols,* Gauthier-Villars, p. 477.
d'Escatha, Y. and Mandel, J. (1971) Profondeur critique d'éboulement d'un souterrain, *Proc. Acad. Sci.,* Paris, **273,** 470–3.
Habib, P., Bernede, J. and Carpentier, L. (1965) Résultats des mesures de contraintes effectuées dans divers souterrains en France, *Annales ITBTP,* June.
Hill, R. (1956) *Mathematical Theory of Plasticity,* Clarendon Press, Oxford, p. 126.
Jaeger, C. (1953/55) Present trends in the design of pressure tunnels and shafts for underground hydroelectric power stations, *Proc. Inst. Civ. Eng.,* March 1953, July 1953 and *Water Power,* February to May 1955.

Jaeger, C. (1958/64) Underground power stations. In *Hydro Electric Engineering Practice* (ed. G. Brown), 2nd edn, vol. 1, pp. 1082–1135.

Jaeger, C. (1979) *Rock Mechanics and Engineering,* 2nd edn, Cambridge University Press, Cambridge, pp. 1–523.

Lombardi, G. (1972) The problem of tunnel supports, *Proc. 3rd Cong. Int. Soc. Rock Mech.,* Denver, Colorado.

Lombardi, G., Electrowatt Engineering Services, and Haertler, A. (1972) St Gotthard road tunnel project; The original concept and design, *Tunnels and Tunnelling,* September to December.

Mandel, J. (1959) Les calculs en matière de pression de terrains, *Revue de l'Industrie Minérale,* January to April, p. 85.

Mantovani, E. (1970) Method for supporting very high rock walls in underground power stations, *Proc. 2nd Cong. Int. Soc. Rock Mech.,* Belgrade, report 6/5.

Rabcewicz, L.V. (1964/65) The new Australian tunnelling method, *Water Power,* no. 16 November/December 1964 to January 1965.

Salençon, J. (1966) Expansion quasi statique d'une cavité à symétrie sphérique ou cylindrique dans un milieu elastoplastique, *Annales des Ponts et Chaussées,* p. 175.

Salençon, J. (1969) Contraction quasi statique d'une cavité sphérique ou cylindrique dans un milieu elastoplastique, *Annales des Ponts et Chausées,* July/August.

Timoshenko, S. and Goodier, J. (1961) *Théorie de l'Elasticité*, Béranger, Paris, p. 87.

INDEX